Asia's Environmental Crisis

Asia's Environmental Crisis

EDITED BY

Michael C. Howard

Routledge
Taylor & Francis Group

LONDON AND NEW YORK

First published 1993 by Westview Press

Published 2018 by Routledge
52 Vanderbilt Avenue, New York, NY 10017
2 Park Square, Milton Park, Abingdon, Oxon OX14 4RN

Routledge is an imprint of the Taylor & Francis Group, an informa business

Library of Congress Cataloging-in-Publication Data
A CIP catalog record for this book is available from the Library of Congress.

ISBN 13: 978-0-367-01201-4 (hbk)
ISBN 13: 978-0-367-16188-0 (pbk)

Contents

Tables and Figures

Tables

Figures

Acknowledgments

I would like to thank Douglas A. Ross and Peter Limqueco for their assistance in the initial planning of this volume and the Donner Canadian Foundation for its financial assistance. Thanks are due to Thea Hinds, Mary Goldie, and Julie Barber-Bowman for their assistance in preparing the papers for publication. Anita Mahoney prepared the final manuscript for publication.

Michael C. Howard

1

Introduction

Michael C. Howard

The struggle of the Penan of Sarawak against the destruction of their forest homeland, Kalimantan's great forest fire, the hundreds of deaths caused by flooding as a result of excessive logging in Thailand and the Philippines, and the air pollution and traffic congestion of Bangkok are only a few of Asia's many environmental problems that have attracted notice around the world.[1] Serious concern about environmental problems in Asia is a relatively recent phenomenon brought about by a rapid deterioration of the environmental quality of the region in recent years in the face of growing population and economic pressures.

Attention has been focused especially on Asia's dwindling forests. Some 5,000 hectares of Southeast Asia's rainforests are destroyed each day, with less than ten percent of this area being replanted. The rate of depletion of these rainforests is fifty percent faster than in Amazonian Brazil. But there are a host of other environmental issues as well. There is considerable concern about the pollution of rivers and farm lands by mines, about air pollution and the contamination of water and soil by industries, about the impact of hydroelectric dams on existing environments, about the impact of biotechnology on forests and farms, and about air pollution and problems of waste disposal in Asia's growing cities.

Assessing responsibility for Asia's environmental problems or finding solutions have not proven to be easy tasks—on technical, economic or political grounds—and it is apparent that situations vary considerably from one setting to another. In the face of such difficulties, there is growing political pressure throughout the region on governments to respond effectively to the environmental crisis. The present volume focuses on the political economy of the environment in Asia. The chapters examine the economic and political forces that have generated the problems, the political efforts to find solutions, and the economic and political contexts of proposed solutions.

Asia's Environmental Problems

How serious are the environmental problems of Asia? For many Asians, concern for the environment remains a secondary consideration, falling well behind the desire to maintain economic growth. They are particularly wary of eco-doom pundits from the developed countries of Western Europe and North America.[2] On the other hand, even to the most casual observer, it is apparent that there is a problem, and that this problem may pose a threat to future economic growth. Thus, Robert Nadelson writes, "the costs of economic development are perhaps nowhere more visible than in China [where] ecological nightmares, such as air pollution and water shortages, already visited upon China threaten to stunt industrial growth."[3] The following is a summary of some of the main environmental problems facing the region.

Forests

Deforestation has received far more attention than any other environmental problem in Asia, and especially in relation to the region's tropical rain forests.[4] The causes and extent of deforestation vary among Asian nations. In tropical areas, an estimated ten percent of deforestation is the direct result of commercial logging, infrastructural development accounts for another ten percent, while twenty-five percent is destroyed by settlers who move into forests after logging and fifty-five percent is due to the demands of agriculture and shifting cultivators. Indirectly, much of this deforestation is linked to population growth and to rural poverty.

In November 1991, flooding on the island of Leyte in the Philippines killed over 3,000 people. The flooding was blamed by many on excessive deforestation and especially on illegal logging allowed by corrupt politicians and military officers. Many environmentalists point to the Philippines as one of the worst examples of deforestation. In 1900, seventy-five percent of the Philippines was covered in virgin forest. By 1990, the area of such forests had been reduced to 2.3 percent of the country (690,000 hectares). Total forest area in the Philippines declined from just over seventeen million hectares in 1934 (fifty-seven percent of the country) to ten million hectares in 1969 (thirty-three percent of the country). Two decades of even more intensive logging reduced the area in 1988 to 6.5 million hectares

(21.5 percent of the country). Recently, the government of the Philippines has increased efforts to conserve the remaining forest (for example, logging in the remaining virgin forest was banned in January 1992),[5] but illegal logging, in particular, remains a serious problem.

Competing with the Philippines for the worst record of deforestation in Southeast Asia is Burma (or Myanmar),[6] where legal and illegal exploitation of the forests, shifting cultivation, and natural fires have resulted in the destruction of some two-thirds of the forest area that existed earlier in the century. Deforestation in Burma became particularly acute after the 1988 coup, when, as phrased by one group opposed to the regime, the government "began to sell Burma's natural resources like fast food."[7] In 1988 Burma still contained an estimated eighty percent of the world's teak forest, but the increased pace of logging, especially by Thai timber companies given concessions by the Burmese government, saw these forests being reduced by over three percent a year by 1990.[8] The situation became sufficiently acute that in early 1993 the Burmese regime threatened to terminate the Thai concessions.[9]

Deforestation is a serious problem throughout the rest of mainland Southeast Asia as well. Prolonged periods of warfare severely damaged forests in Vietnam and, to a lesser extent, Laos and Cambodia. During the Vietnam War, the use of defoliants by American forces resulted in the destruction of some twelve million acres of forest.[10] Since the mid-1970s serious deforestation by farmers and loggers has taken place in these three countries. Widespread deforestation led Thailand (forty-five percent of its forest was destroyed from 1961 to 1985) and then Vietnam and Laos to impose logging bans.[11] Illegal logging, however, remains a serious problem. One result of these bans has been to increase pressure on forests in neighboring countries.[12] In the case of Cambodia, defoliation during the war destroyed a good deal of the forest, but relatively good forest cover remained in several areas. Today, forests in Cambodia's Annamite mountains have been largely destroyed by shifting cultivators. In addition, legal and illegal logging poses a threat to the remaining forest cover. The Khmer Rouge, in particular, granted extensive contracts to Thai logging companies in northwestern Cambodia starting in early 1992.[13] United Nations sources estimated that timber exports reached 1.15 million cubic meters in 1992, a figure that is more than four times the probable sustainable yield.[14]

Forestry practices in the Malaysian portion of Borneo and in Indonesia have been extremely controversial in recent years. An estimated sixty percent of Malaysia's and seventy-four percent of Indonesia's land area are still covered by forest, but pace of deforestation over the past few decades has given rise to concern by many environmentalists. In Indonesia, the focus of attention has been on rainforest degradation, selective logging for pulp and paper, and conversion of forest land to timber plantations, especially in Kalimantan and Irian Jaya.[15] In the case of Kalimantan, by the late 1970s, East Kalimantan alone produced half of Indonesia's timber exports, representing about one-quarter of the country's export earnings. Timber companies cleared huge tracts of forest in the interior of Kalimantan, causing considerable ecological damage. By the early 1980s, three-fourths of the land of Central Kalimantan was set aside for forest production. Most of the timber concessions were held by retired military officers and former government officials. Excessive logging was partially responsible for the "Great Fire" of 1982-83, that resulted in the loss of about twenty percent of East Kalimantan's rainforest. During the remainder of the 1980s, logging operations expanded at an even faster pace. Thus, between 1986 and 1988 annual exports of timber grew from US$1.6 billion to US$3 billion. Further large fires in 1991 drew complaints from neighboring Southeast Asian nations suffering from air pollution caused by the fires.

Deforestation is also a serious problem in the other developing countries of Asia. In China, annual cutting of forest resources surpasses new growth by forty-four percent and an estimated forty-five percent of China's uplands suffer severe soil erosion as a result of deforestation. In Bangladesh, agricultural land use by an expanding population and logging have reduced forest cover to less than sixteen percent of total land area, with no primary rainforest remaining. Nepal's forest cover has been drastically reduced as a result of cutting wood for fuel, clearing forest for agriculture, overgrazing and the use of tree products for fodder, in addition to commercial cutting, and little reforestation.

One important impact of the transformation of Asia's forest lands is the loss of the region's biological diversity. Asia has already lost much of its tree, plant, and animal diversity over the past few decades, and a World Bank survey warns that the region "is likely to lose more of its rich biological diversity than any other region in the world."[16]

Agriculture

Agriculture is an environmental concern in a number of ways. Slash-and-burn agricultural practices in the past may not have harmed the environment so long as population densities were relatively low and market pressures limited, but for the past few decades in many upland regions of Asia these practices have contributed to deforestation and soil erosion as traditional conservation techniques have been abandoned or failed to compensate for greater pressures on the land. The shift to more intensive forms of agriculture also has had a detrimental impact on the environment. As Martin Lewis has remarked, the spread of commercial agriculture has meant that "areas previously marked by sustainable subsistence cultivation soon exhibit such symptoms as rapid soil erosion, deforestation, and pesticide contamination."[17]

As noted above, demand for agricultural land is the most important cause of deforestation in Asia. This can be seen dramatically in countries such as Nepal, where large areas that were formerly forest are now under cultivation.[18] Determining the ways and the extent to which such a transformation leads to environmental degradation is a complex task.[19] Problems arise especially when agricultural practices on newly cultivated land are unable to provide the same level of protection from erosion as forest cover. Lewis documents how pesticides and fertilizers have poisoned farmers and their families living in the Cordillera of the Philippines.[20] Health problems resulting from the ingestion of such contaminants are common in virtually every country in Asia and further demand for increased agricultural production is likely to lead to as much as a seventy-five percent increase in the use of chemical fertilizers in the region by the end of the century.

Ironically, the heavy use of pesticides, rather than reducing the number of pests, in some instances has actually led to a worsening of pest infestation. This has been the case, for example, of broad spectrum insecticides used in Indonesia and Thailand to protect rice crops from brown plant-hoppers. Such insecticides kill not only the pests, but also preditors that normally eat the pests and when pest eggs that have survived the spraying hatch the pest population increases very rapidly. In an effort to deal with the increased pest problem, even more pesticides are used.[21]

Mining

Compared to many other sources of environmental degradation, the impact of mining is relatively concentrated. Of most concern to environmentalists are pollutants from mining that find their way into rivers and streams, thereby being carried beyond the immediate vicinity of the mine itself. Several mines have received international attention as a result of environmental problems. The Ok Tedi mine in Papua New Guinea has been criticized for polluting the Fly River.[22] In the Philippines, large and small mining operations in Luzon's Cordillera and in eastern Mindanao have been blamed for polluting water systems draining into lowland areas. The Freeport mine in Irian Jaya has been accused of polluting waters flowing away from the mine to the southern coast. Tin mines have been held responsible for contributing significantly to the pollution of forty percent of peninsular Malaysia's rivers. Mining has also had a major environmental impact on a number of small islands. In the South Pacific, phosphate mining on islands such as Makatea, Nauru, and Banaba has had a devastating impact.[23] Relatively little attention has been paid to the impact of mining in other Asian countries, including Sri Lanka, India, Thailand, Burma, Vietnam, China, and South Korea.

Energy Consumption

The countries of Asia consume almost twenty-three percent of the world's energy.[24] Perhaps more significantly, this represented an increase in energy consumption of almost sixty percent in a decade, while energy consumption in Western Europe and North America increased very little. For the fastest growing economies, such as those of China and Thailand, and even for some that have grown little, such as the Philippines, energy supplies have not kept pace with demand. Many of Asia's cities, in particular, face serious power shortages. China, for example, increased its capacity by fifty-seven percent during the latter half of the 1990s, but even such an impressive feat has not kept pace with demand.[25] In the case of the Philippines, power shortages have been cited as a major factor in limiting economic growth.[26] Manila provides one of the worst examples of urban power problems, with so-called "brown-outs" lasting up to ten hours a day by 1993.[27] The growth of energy consumption and its continuing increase in the future have important environmental implications in terms of the levels of pollution caused by the use of particular sources. Debate about the relative merits of different energy sources has become more and more intense in Asia in recent years.

Coal accounts for around forty-eight percent of the region's energy and oil thirty-seven percent, with the remainder coming largely from natural gas, hydroelectric dams, and nuclear power plants. Coal is an especially important source of power for India and China, which have large coal reserves themselves. In contrast, almost half of the oil consumed in Asia is imported from outside of the region. Japan imports over seventy percent of its oil from the Middle East. Moreover, oil production within Asia shows little likelihood of increasing significantly. The largest producer, China, can barely meet its own demand, and the second largest producer, Indonesia, may be importing oil within a decade.

The region has large reserves of natural gas, with Indonesia currently accounting for around forty percent of world exports of liquefied natural gas. Expansion of facilities to produce and utilize natural gas have been held back in part by relatively low world prices for the gas and by consumer resistance to price increases that would accompany greater reliance on this fairly clean energy source.

Other potential energy sources that produce less air pollution are hydroelectricity and nuclear power. These sources, however, face considerable opposition on other grounds. Since the 1960s, hydroelectric dams have been criticized on environmental and social grounds. Opponents have been especially concerned with their impact on rural peoples who are forced to relocate because of the dams. Such dams gained a particularly bad reputation in the Philippines during the Marcos years as representing a threat to upland dwelling indigenous peoples of Luzon and Mindanao.[28]

Hydroelectric dams continue to be built throughout Asia and they continue to be subject to considerable criticism. Among the most controversial dam projects in recent years are: the Three Gorges dam project in China, the Tehri and Narmada dam projects in India, the Pergau dam project in northern Malaysia,[29] and the Pak Mun dam project in northeastern Thailand.[30] The Three Gorges dam project entails plans to construct the largest hydropower station and largest ship-lift in the world in the midst of one of the world's scenic wonders. To be constructed in the provinces of Sichuan and Hubei, the project will lead to the resettlement of over one million people, flooding hundreds of towns and villages. The national and international controversy surrounding the proposal is considerable. Thus, in voting for the dam, the governor of Sichuan province acknowledged that the majority of people in his province were opposed.[31]

Activists argue that the Tehri dam in northern Uttar Pradesh threatens the fragile Himalayan ecology and the cultures of the hill-dwelling people of the Garhwal area. The dam will cost an estimated US$2 billion and will be one of the highest dams in the world, and is being built in an area that experienced an earthquake measuring 6.1 on the Richter scale in October 1991. To this, officials respond that the dam will be able to withstand earthquakes measuring over eight on the Richter scale. In addition, the dam will displace around 100,000 people and submerge the most productive farmland in the Garhwal area.

The US$3 billion Narmada Valley Project in Gujarat, Maharashtra, and Madhya Pradesh has elicited even more controversy.[32] This is one of the world's largest hydroelectric and irrigation projects and has been discussed by planners since 1946. Jurisdictional disputes between the three states delayed the project and major construction did not begin until 1989. In the meantime, the project and its backers in the Indian government and the World Bank have come in for widespread criticism by those who claim that the project will cause severe social disruption and environmental damage.

Despite growing public opposition, Asian governments recently have turned increasingly to nuclear energy as a source of power to meet growing demand.[33] Japan, South Korea, and Taiwan, in particular, which rely heavily on imported fossil fuels, are seeking to reduce this reliance through nuclear power. Each of these countries already possesses significant nuclear power generating capacity. Twenty-six percent of Japan's power is produced by nuclear sources (30,385 MW), and the Ministry of International Trade and Industry would like an additional capacity of 40,000 MW by

2010—about one-quarter of this amount is accounted for by units now under construction. South Korea's installed nuclear power accounts for forty-nine percent of capacity (7,232 MW). This is produced by nine units. Two additional units are under construction and there are plans for twenty-four more to be constructed over the next couple of decades. Thirty-eight percent (4,924 MW) of Taiwan's energy comes from nuclear sources, and there are plans to build a new complex capable of producing an additional 2,500 MW.

Other Asian countries have only very limited nuclear capacity, but most anticipate constructing new nuclear facilities. Despite questions about its relative economic efficiency and environmental concerns, the Chinese government is committed to increasing its present modest capacity.[34] At present, there is a small 300 MW unit near Shanghai, and a 1,900 MW two unit complex is nearing completion in Guangdong Province, but there are plans to build at least five more units in the near future. The only existing unit in Southeast Asia is a 620 MW unit in the Philippines that was mothballed in 1986. In the face of a serious energy crisis, the Philippines government is negotiating to have the unit opened by 1995-96. Elsewhere in the region, Indonesia,[35] Malaysia, and Thailand[36] have indicated that they are considering construction of nuclear units during the 1990s. In South Asia, both India and Pakistan have small nuclear reactors in operation. India has seven reactors, but their total capacity is only 1,374 MW. India has a target of increasing its nuclear capacity to 6,000 MW, but this may be overly ambitious. Plans to construct a large plant in Tamil Naidu were recently canceled following the collapse of the Soviet Union. Pakistan has one small 140 MW reactor in Karachi, and the Chinese are about to build a 300 MW unit with talks underway for a second unit, but a lack of capital makes further development uncertain.

Urbanization and Industrialization

Asia is the home of many of the world's largest and most rapidly growing cities, and most of these cities face serious environmental problems.[37] Thus, more of China's cities exceed World Health Organization guidelines for air quality than any other country. South Korea's cities, likewise, suffer from high levels of air pollution caused by motor vehicles, coal heating, and factories. Even in Japan, although authorities have tightened regulations in an effort to reduce pollution, levels of air and other forms of pollution in urban areas continue to increase. In mid-1985, Japanese government authorities estimated that some 100,000 people suffered from pollution-related diseases.[38] In addition, Japanese city-dwellers produced 36.7 percent more garbage in 1992 than they had in 1982 and existing land fills are expected to have reached capacity by 1998.[39]

The cities of Southeast Asia also face serious pollution problems as a result of rapid population growth and industrialization that has far outstripped urban infrastructures. With its large number of international visitors, the environmental problems of Bangkok have attracted particularly widespread attention. Thailand's rapid economic expansion and industrialization, and the fact that much of it is concentrated in and around Bangkok, has resulted in serious problems from airborne dust and lead pollution in the Bangkok area. The largest concentration of factories are in the province of Samut Prakan, where the provincial governor noted that half of the province's 4,000 factories "cause annoyance to people living nearby" and "31 cause serious pollution problems."[40] Some 500 new vehicles a day are added to Bangkok's streets. Urban infrastructure remains woefully inadequate. For example, only two percent of Bangkok's nine million inhabitants are served by sewage facilities. Airborne pollution is estimated to be responsible for as many as 1,400 deaths per year.[41] Industrial pollution is threatening to spread elsewhere in Thailand as a result of government initiatives to lessen the concentration of industry in the Bangkok area by offering incentives to build new industries elsewhere in the country. Sharp increases in air pollution from urban and industrial areas threaten much of the Malay peninsula, with the problem being especially serious in Kuala Lumpur and the Klang valley.[42]

Even in the South Pacific, urban centres such as Fiji's Suva and Majuro in the Marshall Islands are increasingly polluted.[43]

Water

A report by the World Bank cited contaminated water as a top priority for developing countries.[44] Much of the pollution is caused by human waste, but improper disposal of chemicals and other hazardous wastes from industry and farming is also an important source of pollution. Obtaining safe water is a growing problem throughout urban and rural areas of Asia as a result of growing populations, industrialization, and the use of polluting chemicals in agriculture and other rural industries. In Japan, acid rain has led to significant degradation of water quality, threatening aquatic life and humans. In South Korea, despite economic growth, almost half of the population does not have piped water, and much of the water available is polluted as a result of the lack of sewage treatment facilities and discharges from industries. Pollution from the use of chemicals by Korean farmers is also a serious problem in rural areas. In China, only about forty percent of the population has access to potable water, and water pollution is contaminating an increasing amount of agricultural land and marine products. A major source of water pollution in China are the state-run factories.

Its rivers and streams have been central to life in Southeast Asia for thousands of years and, until recent decades, relatively low population densities and low levels of industrialization meant that these waterways were relatively clean. Population and economic growth has resulted in the rapid deterioration of water quality. The situation is especially serious in urban areas, such as those of Jakarta, Manila, and Bangkok. In the case of Bangkok, only around two percent of its inhabitants are served by sewage facilities. The Chao Phraya River around Bangkok is seriously polluted, with oxygen levels in some areas close to zero. In rural areas intensive use of chemicals has become a major source of water pollution. In addition, in parts of the Philippines chemical pollution from small-scale gold mining has led to extensive water pollution in Luzon and Mindanao. Likewise, Malaysia's tin mines have heavily polluted many of its rivers. Pollution of the water has also resulted from the growth of tourist resorts, especially in Thailand. The resort of Pattaya, as noted by one author, "has earned notoriety as one of the most disgusting bathing spots in the world."[45]

The lack of clean water is a particularly acute problem in South Asia. India's rivers and streams are highly polluted and most people do not have access to clean water. The major cause of water pollution is untreated sewage and other nonindustrial wastes (four times as much as industrial waste), and only a 219 communities out of the country's more than 3,000 have even partial sewage treatment. The seriousness of the problem is highlighted by the fact that waterborne diseases account for two-thirds of all diseases in India. Large sections of the Yamuna River, with a catchment area that covers 10.7 percent of India, are heavily polluted. Some of this pollution is caused by industrial pollution, but most is from untreated sewage.[46] A similar situation prevails in the other countries of the region. In Bangladesh, where ninety-seven percent of rural dwellers have no sanitation, the problem is made even worse by floods which spread polluted water across areas that are used for fishing and rice cultivation. In Bhutan, only about twenty-five percent of the rural population has access to safe drinking water and only about seven percent of the population has access to sanitation services.

Aquaculture and Coastal Regions

Asia's coasts have been subjected to environmental threats from a variety of sources, including aquacultural development, the destruction of mangroves, and the spilling of contaminants. Aquaculture has been an important activity for an extremely long time in East Asia and Southeast Asia. During the past couple of decades efforts have been made to expand and intensify aquaculture throughout these parts of Asia as a means of promoting rural development and earning foreign exchange. The results of this initiative in economic and environmental terms have been mixed and often unclear.[47] In terms of environmental problems, one study of the impact of the rapid expansion of shrimp farming in southern Thailand, for example, provides a list of accompanying environmental problems, including "degraded water quality from waste water discharges, salt water intrusion into rice paddies or drinking water supplies and exacerbated flood hazards."[48] The study notes that pollution and related problems are sufficiently serious as to threaten the viability of shrimp farming in the area in the near future.

In the not too distant past, much of Southeast Asia's coastline was covered in mangrove. As the mangroves have disappeared, ecologists have pointed to their environmental importance and to the negative impact of their loss.[49] In the Philippines, natural disasters combined with the effects of inland deforestation and over exploitation of coastal resources have led to serious deterioration of mangroves and other coastal resources.[50] The spraying of herbicides during the war in Vietnam and subsequent clearing of areas for aquaculture have destroyed much of Vietnam's mangroves. Much of the mangrove that once covered Cambodia's coast is now gone. Siltation, tin mining, and other factors have combined to destroy most of peninsular Malaysia's mangroves, especially on the west coast. In Thailand, aquaculture and mining along with other activities have led to severe deterioration of mangroves.

In southern India and Sri Lanka, mining of coral and siltation caused by inland agricultural and forestry practices have severely damaged mangroves.

Pollution resulting from oil spills is another source of deterioration of coastal environments, especially in Pakistan and peninsular Malaysia. An Indonesian scientist has blamed increasing levels of coastal and sea pollution on offshore gas and oil drilling in addition to poor agricultural practices, mismanagement of the forests, and improper waste disposal in urban areas.[51]

Wildlife

The loss of biological diversity in Asia was mentioned in the section on forestry. The loss is particularly acute for Asia's wildlife, which has fallen victim to the forces of population and economic growth and environmental degradation. Deforestation in inland areas and widespread destruction of mangroves along the coasts has had a serious impact on wildlife. In South Korea, for example, deforestation, the use of pesticides and fertilizers, and other activities have led to the loss of over 100 animal species in recent years. Deforestation in Cambodia has resulted in a loss of an estimated three-fourths of its wildlife habitat. Densely populated Bangladesh has very little wildlife habitant left.

Worldwide efforts to promote wildlife conservation were given a boost in 1980 with the launching of the World Conservation Strategy. The emphasis of the strategy was sustainable use of living natural resources and conservation of biological diversity. Efforts to protect habitat areas and strategies for doing so vary from country to country in Asia. Sri Lanka has a long tradition of habitat conservation, with over forty percent of its land area protected in some way, but poaching and the civil war pose a serious problem for its wildlife. In Southeast Asia, Thailand and Indonesia have each set aside between eight and nine percent of their land as nature reserves, parks, and other types of protected area (47,000 and 141,000 square kilometers respectively. The figures for Malaysia and Vietnam are three percent each (11,000 and 9,000 square kilometers respectively), and the Philippines only two percent (5,000 square kilometers). In east Asia, Japan and South Korea have set aside six percent each (24,000 and 6,000 square kilometers respectively), and China a mere one percent (79,000 square kilometers).[52]

Throughout Asia, poaching and illegal incursions for agricultural purposes remain a problem, with poaching of endangered species for sale illegally a widespread concern. In addition, wildlife conservation efforts have often generated conflicts with poor rural farmers who generally are asked to give up access to land for the sake of creating wildlife refuges.[53] Effective enforcement of regulations is beyond the present capability of most governments in the region, much as is the case with enforcement of forestry regulations in general, and there is growing awareness of the need for community-based conservation efforts.

Animal species are also disappearing in response to a worldwide market for exotic animals as pets, food, and the like. Many of these animals are smuggled to developed countries in Europe and North America, but there are markets for them in Asian countries as well. The more developed Asian countries, such as Japan, are often pointed out as being particularly guilty, but the problem exists throughout the region.

Leisure Activities

Tourism and golf have also figured as environmental concerns in Asia. Philip Dearden has noted that tourism "can be a major engine of economic growth and, compared to some industries...can be relatively environmentally benign."[54] But, as Dearden illustrates in the case of Thailand, unregulated tourist development can have a very adverse impact on the environment. Thailand's beach resorts, in particular, have received considerable attention because of extent to which they have dumped raw sewage into the waters surrounding them. Fecal contamination in Pattaya Bay, for example, exceeds international standards by more than ten times.[55] Environmental concern has also focused on tourist developments around relatively pristine coral reef areas, such as Pulau Redang off the east coast of Malaysia,[56] which are extremely vulnerable to human activities.

As golf courses have spread across the globe, they have come in for increasing criticism from environmentalists. Criticism has focused on the use of chemicals to maintain the grass which cause pollution and harm wildlife and on their demand for water. The Global Network for Anti-Golf Course Action, based in Tokyo, is especially critical of Japanese companies building golf courses overseas with little regard for the environmental consequences.[57] Vietnamese authorities have promoted golf courses as a means of attracting foreign tourists. One project by a Taiwanese company was abandoned only after the company created an ecological disaster by clearing 234 hectares of trees and other growth in a seaside park. In another case, a joint venture between three Vietnamese companies and a Taiwanese company was granted permission to build a resort with two golf courses in the center of a green belt, that includes the Thu Duc forest, established near Ho Chi Minh City in the early 1980s.[58] In addition to efforts to stop construction of golf courses, critics have called for efforts to construct and maintain golf courses in ways that are environmentally more sound.

Corruption has been a factor contributing to the spread of leisure facilities into environmentally sensitive and reserved areas. Golf courses and resort complexes commonly appear in such areas either without official permission or after having permission granted under suspicious circumstances. Corruption involves project developers seeking to bribe officials as well as running bribery rackets to allow developers to build in areas where permission is difficult to obtain. Sometimes, those involved are caught and the projects stopped or removed, but this does not appear to happen often, especially when the developers are politically influential.

Global Warming

The extent of concern and of debate over so-called global warming and the resultant prospects for sea level rises and climatic changes were highlighted at the 1992 Earth Summit in Rio and the Climate Change Convention. The United Nations-sponsored Intergovernmental Panel on Climate Change, a few months earlier, reduced earlier warnings concerning the pace of global warming, but cautioned that the threat was still considered serious. While debate continues concerning the threat of global warming, several countries in the Asia-Pacific region have taken steps to examine the potential damage.

A study of the possible impact of global warming on Southeast Asia, raised the prospect of significant declines in agricultural production and of flooding of important coastal areas.[59] The study warned, in the case of Thailand, that 5,000 square kilometers of land could be flooded as a result of a one meter rise in the sea level, and that such flooding would threaten Bangkok.

Global Warming has also been of concern to several South Pacific Island nations. The issue was raised at the 1988 South Pacific Forum meeting in Tonga, at which the Australian government offered to fund setting up monitoring stations in the region. The Australian offer was scoffed at by Prime Minister Mara of Fiji as an exaggeration, but it was taken seriously by other Pacific island states, especially Kiribati, the Marshall Islands, Tokelau, and Tuvalu, which are composed entirely of low-lying atolls. The problem was highlighted further at the 1989 South Pacific Forum meeting held in the atoll-nation of Kiribati, which would largely disappear should the sea level rise one meter.[60]

The countries of East, Southeast, and South Asia themselves contribute roughly twenty-six percent of greenhouse gas (carbon dioxide, methane, and CFCs) emissions each year, with China, India, Japan, and Indonesia being responsible for the largest portion of this (18.7 percent).[61] On a per capita basis, however, these three countries do not rank especially high among the nations of the world (ranking 104th, 113th, forty-fourth, and seventieth respectively). Nevertheless, population growth and the rapid pace of industrialization in much of Asia raises concerns that the region may come to contribute a much greater share of greenhouse gas emissions in the near future.

The Environmental Movement in Asia

Until the 1960s, by and large, regulating the environmental impact of human activity on the environment in Asia was left to informal conservation practices and, more formally, to relatively small and weak government ministries and departments, such as those concerned with forestry and mining.[62] By the late 1960s and early 1970s, there was limited but growing public concern that existing institutions and policies were inadequate to deal with the environmental impact of postwar population and economic growth in Asia as well as with the power of the industrialists and others who opposed environmental regulation. Environmental concerns were articulated through non-governmental organizations by naturalists, such as bird-watchers, upset over the impact of environmental degradation on wild plants and animals and by those troubled by problems faced by rural-dwellers in the face of harmful commercial activities such as logging and hydroelectric dam construction, as exemplified by opposition to the proposed Chico River dams in the Philippines.

As Hasin Ali has noted in the case of several Southeast Asian countries, a variety of non-governmental organizations emerged in the 1960s and 1970s within the context of political authoritarianism as a "vehicle for expressing political dissent and socio-economic grievances."[63] Such organizations often sought to respond to problems caused by highly centralized developmental initiatives in rural areas. Directly, or by implication, they presented a challenge to authoritarian politics and to an associated top-down model of development. However, these non-governmental organizations voiced environmental concerns to varying degrees. Among the earliest groups, organizations, or movements to focus on environmental issues were the non-violent Chipko movement in India, which started in 1973 among low caste people in the Himalayas,[64] and two Malaysian organizations, the Sahabat Alam Malaysia and the Environmental Protection Society Malaysia, which were also founded in the 1970s.[65] These organizations paid particular attention to the effects of deforestation, with those in Malaysia also being concerned with environmental issues related to industrial pollution and mining.

Since the 1970s there have been a growing number of environmental groups throughout Asia. In the Philippines, for example, the Haribon Foundation for the Conservation of Natural Resources, founded in 1972 as a bird-watching club, became a vocal environmental group in the early 1980s. In Indonesia, WALHI (the Indonesian Environmental Forum) was founded in 1980 as an umbrella organization for naturalists, environmentalists, academics, and others. In Thailand, the Project for Ecological Recovery was formed in the early 1980s, and since then other groups have been formed.

Environmental movements have not developed uniformly throughout the region and it is difficult to assess how effective they have been. In the case of Japan, Brendan Barrett and Riki Therivel note that while groups have rallied around specific local environmental issues, a national movement has failed to develop.[66] Such local pressure contributed to national and local governments establishing environmental impact assessment procedures in the early 1970s, but in the late 1970s more powerful ministries reduced the power of the environment agency and since then its power has been further eroded. On the other hand, opposition by non-governmental organizations in Thailand to the Pak Mun dam led the World Bank to withhold a US$22 million loan for the project.[67] This was the first time that the World Bank took such a step as a result of pressure from non-governmental organizations and foreshadowed problems in obtaining funding for other dams in Asia, such as the Narmada Valley Project in India, and a much more cautious approach to dam construction in Thailand.

Non-governmental organizations have been an especially important feature of recent political life in the Philippines.[68] Such organizations have been very active in environmental affairs as critics and in directly applied activities. Following inauguration of the Community Forestation Program in 1990, for example, the government awarded a number of non-governmental organizations contracts to organize and train communities.

The environmental movement in Asia is part of an international movement and has been influenced by environmentalists elsewhere. International support has been important to many environmental groups in Asia, but it has also left them open to criticism, especially from governments sensitive to challenges of existing developmental models and policies. Such critics have argued that many of the environmental policies being advocated represent a new form of neocolonialism as those from the developed world seek to maintain their dominance of the developing world. The Malaysian government has emerged as an important critic of the environmental movement along these lines. The Malaysian government has sought to defend its development policies in relation to forestry as being in the best interests the people of Malaysia, while accusing its critics of paternalism, neocolonialism, and of being political rather than scientific.

The Malaysian government has been especially critical of foreign environmentalists, who it has accused of improperly interfering in Malaysia's internal affairs. Thus, ten activists from Robin Wood of Germany, Earth First of the United States, and the Society for Threatened People of Switzerland were arrested in Sarawak in mid-1991 after chaining themselves to loading cranes on barges carrying logs following weeks of protests by members of the Penan and Kelabi tribes against logging activities. The chief minister of Sarawak stated that these environmentalists, who had entered the country as tourists, had no right to engage in such activities.[69] The Malaysian government has argued that it monitors logging activities sufficiently to prevent abuses and that it plans to resettle the tribal people in modern homes and to integrate them into modern Malay society. Concerning their responsiveness to environmental issues, Malaysian officials point to the fact that, in response to recommendations in 1990 by a team from the International Tropical Timber Organisation, it has set about to reduce timber harvesting from forest estates in Sarawak by twenty-five percent over two years.[70]

Indonesian officials have also been critical of what they view as irresponsible criticisms by environmentalists and have defended their policies as being responsive to environmental problems. The Minister of

Forestry Hasjrul Harahap is quoted as saying that "Indonesian's forestry activities are an open book showing sustainable development."[71] Nevertheless, while the Indonesian government has taken steps to regulate the timber industry in recent years, private interests within the industry have continued to exhibit a high degree of independence and illegal logging remains widespread. In mid-1992, for example, clans in the Sorong area of Irian Jaya complained that the government was ineffective in dealing with fourteen timber companies that were accused of illegally cutting trees of some 280,000 hectares of forest land. A 1990 study sponsored by the Food and Agricultural Organisation questioned the sustainability of logging practices in Indonesia. The study reported current log production was around thirty-seven million cubic meters, while a sustainable level would be about twenty-five million cubic meters.[72]

One particularly heated controversy involving national and foreign non-governmental organizations, private national and corporate interests, and the Indonesian government was that surrounding the proposed Scott Paper and Astra International eucalyptus plantation/woodchipping and pulp mill project in Irian Jaya. The project was to have cost US$645 million, making it the largest foreign investment in Indonesia outside of the oil and gas industry. The project was especially important to the government of Irian Jaya, eager for outside development funds. Shortly after the plan was announced, foreign and Java-based non-governmental organizations launched a worldwide campaign to force Scott Paper to withdraw. When Scott did decide to pull out in late 1989, as noted by a publication of one of the organizations involved in the campaign, the company's decision was "cheered by environmental and human rights groups worldwide."[73]

The Indonesian non-governmental organization SKEPHI was especially vocal in its call not only for Scott Paper to withdraw, but, more broadly, for the government to pursue an alternative development policy for all of Irian Jaya based on ecological sustainability. SKEPHI referred to Irian Jaya as "the last mystery of the world," and felt that this unique place needed special consideration. The position of many of those within Irian Jaya itself tended to be more moderate. Some local non-governmental organizations in Irian Jaya were not so keen on the anti-Scott campaign, and, in fact, they had not been consulted before the campaign was launched. Those involved in the campaign had little time to celebrate when a short time later it appeared as if Taiwanese or other Asian interests, who were clearly much less environmentally conscious than Scott, might take Scott's place. When Astra pulled out of the project in September 1992, those wishing to make the last mystery of the world a model of small-scale community development were pleased, but more moderate environmentalists in Irian Jaya and elsewhere in Indonesia were less enthusiastic with what could also be seen as a setback for developmental efforts in Irian Jaya.

Scott's withdrawal from the project contributed to a backlash against some of the more vocal non-governmental organizations in Indonesia in late 1989, especially those with extensive foreign links. Minister of Forestry Hasjrul Harahap accused these non-governmental organizations of offering no viable alternative forms of development, while State Secretary General Moerdiono warned them against using environmental issues to camouflage destabilizing political activities. The issue of international links arose again as relations between the Netherlands and Indonesia deteriorated following the November 1991 massacre in Timor Timur. When Indonesia suspended Dutch official assistance this affected a number of non-governmental organizations which were being supported by the Dutch financially, accusing them of being unduly influenced by foreign interests.

Many of the environmental non-governmental organizations in Asia are part of a ‘general political movement that is opposed to existing authoritarian tendencies and favors promoting more direct forms of democratic participation. This is not to say, however, that there is uniformity of political purpose within the environmental movement. Some organizations are politically relatively neutral, carefully working within the existing system to promote environmental protection, usually in a relatively limited way. Other organizations are politically more assertive in their advocacy of environmental and democratic causes, but do not espouse a worldview that is radically different from many government planners, politicians, and those in the private sector. Such organizations support development, so long as it is environmentally sustainable. They differ from the first group primarily in the degree of their skepticism concerning the ability of government and business interests to act responsibly and in their desire for more thorough democratic reform. Then there are those who advocate a more radical environmentalism. These organizations generally advocate curtailing development of the sort advocated by the region's governments and support an idealized rural lifestyle of small-scale, technologically simple production and approximate equality of wealth and power. While media and government attention regarding Sarawak's rainforests, the Scott-Astra pulp and paper project, and the like usually focuses on the more radical organizations, it is important to recognize that their views do not reflect the totality of environmental thinking within Asia and that the region's environmental movement is very heterogeneous.

Proposed Solutions

The heterogeneity of environmentalism in Asia can be seen in relation to proposed solutions to the region's environmental problems. On the one hand, there are those who believe that modern technology and centralized management are capable of dealing with whatever environmental problems might arise as a result of continued economic growth. Regulation, from this perspective, is largely a series of directives telling people what to do and what not to do with penalties for failure to comply. On the other hand, there are those who support community-based management and who eschew modern technologies in favor of more traditional methods of conservation and production.

In recent years there have been many innovations to allow for more efficient and less polluting production and use of resources. For example, there have been important technological advances in the pulp and paper industry that allow utilization of resources other than the trees currently used, thereby taking some pressure off of existing forests, and a reduction of pollution by the industrial plants. The mining industry has seen a number of important innovations in its ability to handle wastes, although little of this technology has found its way into Asia so far. Bioleaching is one example, involving the use, for instance, of *thiobacillus ferro-oxidans*, a naturally-occurring bacteria, to liberate metals such as copper, gold, and uranium.[74] Another example is the use of dry tailings systems, such as the one installed by Placer Dome at the La Coipa gold mine in Chile, which serves to optimize water use and minimizes risks of contamination.[75] In the energy field, experiments in several areas hold out hope for changes to less polluting methods. Among the possible innovations in energy use are advances in solar power technology,[76] cold fusion, the use of less polluting fuels in vehicle engines, pressurized biomass gassification,[77] and new means of burning coal such as fluidized bed combustion and the use of integrated gas combined cycle technology.[78] There are also a growing number of efforts to conserve energy and to use it more efficiently, such as the production of more energy-efficient refrigerators.[79]

Asia produces much of the world's textiles and, as a result, is also the consumer of the bulk of dyestuffs, which produce high levels of air and water pollution as well as other hazards for those using them. Java provides an example of some of the worst pollution in the dye and textile industries, but it is also the site of important efforts to overcome these problems. A Ciba-Geigy plant near Jakarta is notable for its innovative methods of waste water and dye recycling.[80] Experiments are also being conducted in Jakarta to allow for the use of organic dyes in the batik industry on a large commercial scale.

While it is possible to point to some instances where new, environmentally more sound technology has been employed in recent years, there are far more instances where this is not the case.[81] Moreover, there remain important political and economic questions concerning control of the technology and access to it. This is especially a concern in the case of biotechnology, but in other areas as well transnational corporations from developed Western countries dominate much of this new technology, raising fears of a new dependency.[82]

Throughout the region, government's have made efforts to improve management of the environment. Most governments in Asia now have environmental ministries and some kind of requirement for assessment of the environmental impact of activities prior to their being approved. In some instances, such initiatives have occurred only very recently and environmentalists wonder if they are adequate for the task. Taiwan established its Environmental Protection Administration in 1987, but so far has made little headway in trying to clean up after decades of rapid, virtually uncontrolled growth.[83] Hong Kong established an Environmental Protection Department, and has taken some steps to lessen air, noise, and water pollution. Thus, noise controls were put in place in 1988, the use of high-sulfur fuels in factories was banned in 1990, and a US$2.5 billion ten-year clean-up plan launched in mid-1989. Despite these efforts, as noted by one commentator, "Hongkong still looks, smells and sounds like a highly polluted city and will continue to do so for many years."[84]

There is perhaps no better example of an area where Asian governments have clearly failed to act than waste treatment. Although widely recognized as a serious problem for decades, governments have provided only limited resources for dealing with the problem. This situation has improved slightly in recent years. Governments in Malaysia, Thailand, Indonesia, and Hong Kong, for example, awarded contracts to private firms to install treatment plants in 1993,[85] but critics feel that such efforts still fall far short of what is needed.

As with the example of Japan discussed above, these environmental ministries usually are among the least powerful in the governments. This is quite clearly the case in India, for example, where Manekha Gandhi was named Minister of State for Environment and Forests, being the only member of the National Front who wanted the portfolio in a government planning to go ahead with such controversial undertakings as the Narmada project.

The government of Thailand has improved its management of the environment considerably in recent years, but efforts to improve waste disposal, air quality, and forest protection are still viewed as far from adequate even by moderate environmentalists.[86] While Thailand's new environmental legislation enacted by the government of Prime Minister Chuan Leekpai in 1992 "has been lauded around the world for being sweeping and progressive," as journalist Paul Handley notes, "as it turns out, this is part of the problem. The new approach is so advanced that the country's sleepy, much-ignored 17-year-old environment agency has not yet got the manpower or the skills to work with these new tools."[87] In addition, the establishment of a new environmental think-tank in 1993, the Thailand Environmental Institute, drew attention to sharp differences between those focusing on environmental degradation and technical solutions and those more concerned with social impacts and social solutions.[88] While the existing bureaucracy favored the former approach, those behind the new agency felt that more attention needed to be paid to the latter. Those behind the new agency also advocated a greater emphasis on the role of non-governmental organizations and popular participation rather than highly centralized technical planning. Similar issues are being debated throughout Asia and reflect broader pressures for changes in the ways governments make decisions.

Even well intentioned government efforts to better monitor environmental regulations often are inadequate to stop illegal activities. Those seeking to stop illegal logging in countries such as Thailand or Indonesia frequently find their efforts thwarted by the political connections or by the superior firepower of those involved. At other times, government efforts themselves may be suspect. For example, when the Indonesian government awarded a forestry-inspection contract to the consulting firm of Reid, Collins of Vancouver to inspect logging concessions in three provinces, critics noted that the firm's Indonesian partner was part of the Bimantara group, a diversified conglomerate controlled by Bambang Trihatmodjo, President Suharto's second son.[89]

Criticisms of the effectiveness of government regulatory efforts have focused on problems in the forestry sector and this is the area that has received the most attention by advocates of community-based management. Critics argue that the "punishment rather than prevention approach"[90] does not appear to be a particularly effective or efficient way of managing forest (or other) resources. To begin with, it can be argued that there will never be sufficient government personnel effectively to handle forestry problems on their own. Protection of forests is only really possible, so the argument continues, when those who live in and around the forests, and who traditionally managed forest resources on their own, are made responsible for management of forest resources rather than being treated as potential destroyers of forests.[91]

Community-based management, its proponents contend, also helps to overcome the preservation versus development problem. Based on an extensive study of community efforts to manage and protect forest resources in northern Thailand, Anan Ganjanapan finds that "the experiences of such community forestry demonstrates clearly that humans and forests can live together in a meaningful way that allows for development and conservation."[92] Support for community-based forest management has also come from an increasing number of multilateral and bilateral development agencies which have recognized that poverty and landlessness are crucial issues that need to be confronted if forest management is to be improved and view community-based forestry as a means of addressing the dual goals of development and conservation. Years of pressure from academics and non-governmental organizations and greater government support for decentralization led Thailand's forestry department in early 1993 tentatively to back the idea of community-based forestry.[93] In the past, the department had favored technical and highly centralized approaches that included the removal of villagers from forest areas under its jurisdiction.

Advocates of community-based forestry, however, have found themselves in a very delicate position since their position challenges important interests and runs counter to the tradition of strong central authority that is to be found throughout much of Asia. Centralized governments of Asia are generally not sympathetic to allocating too much power to local communities, especially in strategically sensitive areas and where valuable resources are concerned.[94] This problem is illustrated, for example, by debate within Indonesia over control of resource-based development in Irian Jaya, where locals have been concerned with forestry, mining, and tourism activities by outside interests. In reference to the conflict over the Narmada Valley Project in India, M. Shugenne argues that "the most revealing facet of the latest cycle of mega irrigation and dam projects is the complete lack of information at the grassroots."[95] Even in the Philippines, where provisions for community-based forestry management have been in place since 1982, only a handful of so-called Community Stewardship Agreements have been awarded.[96] The government of Papua New Guinea, where decentralization has been a more salient issue than in most Asian countries, has sought to devise more equitable means of distributing decision-making and the wealth derived from mining, forestry, and other resource-based industries at the local, provincial, and national levels. As exemplified by the Bougainville crisis, however, progress in this regard has proven difficult and relations at these various levels remain strained.

Logistical problems are often cited by opponents of community-based resource development. The complexities and ambiguities of land issues in many tribal areas is a commonly mentioned reason among those seeking to develop resource-based projects for relying on more centralized and clearly delineated decision-making models. An example of just how hard implementing a community-based project can be was demonstrated by the problems faced by the New Zealand-funded Bukidnon Industrial Plantation Project in Mindanao, which was confronted with an array of sometimes very questionable land claims by local tribals.[97] Advocates of community-based development counter that such obstacles are not really so great and that they can be resolved by incorporating local people in the decision-making processes.

A more fundamental issue concerns the idealized view of local, especially indigenous communities presented by supporters of community-based development. Some critics of the approach in Thailand point to extent to which traditional forms of leadership and resource management in rural communities have already broken down and to the corruption of many local authorities. In regard to this last point, local chiefs and other political figures in the South Pacific have proven quite susceptible to bribes and other temptations and to the schemes of unscrupulous developers where the illegal or unsound use of natural resources is concerned. Corruption and self-interest are characteristics found across cultures at all levels of government. Such examples do not necessarily negate the possibility of establishing sound community-based forestry practices, but they do provide a cautionary note.

An additional question concerns the applicability of the community-based development model developed for forestry for other industries such as tourism or mining. On economic and environmental grounds, some economists argue that large-scale highly concentrated tourism is superior to more diffused forms of small-scale tourist development such as that associated with the approximately 100,000 trekkers who visit northern Thailand every year. In the case of mining, unlike forestry, there are rarely desirable alternatives to the methods already available to large mining enterprises for extracting mineral resources. Small-scale mining operations, on the whole, have a terrible environmental record and often reduce the prospects for local sustainable development. Experience has also pointed to the importance of economies of scale in the mining industry and to the advantages that transnational corporations often have in regard to processing, refining, and marketing. The implication is that local people generally have limited options. In effect, they can seek to block the development of a mine altogether or they can attempt to make the large-scale mining enterprise more socially responsible in its hiring, working, operational, and clean-up and restoration practices.[98]

The state of the environment and proposed solutions to perceived environmental problems have become an integral part of political and economic debate throughout Asia. Environmental issues in Asia and the arguments surrounding them are rapidly evolving in response to an array of local, national, regional, and international factors. It is also possible to view the debate as becoming more complex as it moves beyond some of the simple notions that characterized it at the outset. In particular, while some commentators on Asian environmental problems still view the situation simplistically, most are all too aware of the need to ground their analysis in the complicated political, economic, social, and cultural context of Asia rather than simply importing ideas from elsewhere.

The Contributions

The twelve chapters that follow analyze a range of environmental problems in a variety of settings. The first two chapters focus on questions relating to the environment and indigenous rights. G. Peter Penz raises ethical questions about environmental justice in relation to conflicts over land between the states of Bangladesh and Indonesia and indigenous peoples in the Hill Tracts of Bangladesh and the Indonesian province of Irian Jaya. Susana B.C. Devalle analyzes the struggle between indigenous peoples and those seeking to develop the forests of Jharkhand in the Indian state of Bihar.

The next two chapters examine mining, with some attention to its relevance to indigenous peoples. The first of these, by Michael Howard, looks at the debate over the relative impact of small-scale versus large-scale mining in Southeast Asia. Many environmentalists have championed the cause of small-scale mining, in part because of its supposed benefit to indigenous peoples, yet the issue is a complex one and the record of the environmental impact of small-scale miners is far from benign. The chapter by Donna Winslow examines the environmental impact of mining (primarily nickel mining) on New Caledonia. Her chapter also relates mining to issues of indigenous rights, since New Caledonia has been embroiled for many years in an independence struggle by its indigenous population.

Commercial tree cultivation has replaced natural forest throughout much of Southeast Asia. Alec Gordon discusses commercial rubber cultivation in Southeast Asia by small-scale producers, and argues that rubber cultivation provides a sustainable alternative to natural forest. Gordon also raises the issue of the impact of deforestation on wildlife and points to how rubber can serve as a habitat especially for birds. Apichai Puntasen, Somboon Siriprachai, and Chaiyuth Punyasavatsut, in their chapter on eucalyptus growing in Thailand, in contrast, argue that eucalyptus has a very damaging impact on the environment. The Thai eucalyptus industry has expanded, nevertheless, the authors argue, largely through the connivance of individuals in government and the private sector lured by potential profits.

Biotechnology is at the heart of much of the debate concerning sustainable development in agriculture and forestry as well as concerning problems of biodiversity. Nagesh Kumar argues that while biotechnology has assisted in increasing production, there are important questions about the sustainability of this growth and about its social and economic impact. Biotechnology is heavily dependent upon non-renewable resources, its application has led to land degradation in some instances and to a loss of genetic diversity. Moreover, biotechnology can be seen as contributing to the concentration of wealth and to greater inequality. Kumar expresses particular concern about the extent to which biotechnology research and its findings are controlled by transnational corporations. This raises the specter of increased dependency of agriculture in developing countries and an even wider technological gap between the rich and poor nations. Such a situation makes it extremely important for the developing countries of Asia to ensure that the application of biotechnology and biotechnology research are more in keeping with their long-term interests.

The final five chapters focus on environmental issues in particular Asian countries. The chapter by Mark McDowell examines China's energy strategies in relation to environmental constraints on the one hand and the pursuit of economic growth on the other hand. China faces a major energy problem and efforts to overcome it have generated considerable controversy. The chapter examines one of the most controversial initiatives, the Three Gorges Dam Project, which is to include the installation of the world's largest hydroelectric generating project. McDowell argues that, despite problems associated with hydroelectric projects, the benefits of hydroelectricity outweigh the drawbacks in light of the options available— such as greater reliance on coal or nuclear power—but that projects of a smaller scale are preferable to the grandiose Three Gorges Dam Project. The nuclear option is viewed as especially worrisome in light of China's poor industrial safety record.

The next two chapters are on the Philippines. The first, by Rene Ofreneo, focuses on an issue that is of relevance for the region as a whole—the relationship between Japanese investment and environmental degradation in the Philippines. Japanese capital has been particularly active investing in resource industries in the Philippines and elsewhere in Asia to provide Japan with natural resources. The result, all too often, has been widespread environmental damage that has generated a good deal of controversy. The chapter by Rosalinda Pineda-Ofreneo discusses another issue that is of importance for other Asian countries besides the Philippines—national indebtedness and the environment. Pineda-Ofreneo argues that the debt-trap within which the Philippines finds itself contributes to environmental degradation in terms of the pressure it places on the government to allow private interests to exploit natural resources in ways that are not environmentally sound and because the related poverty pushes those who are poor to over-exploit the environment.

Dong-Ho Shin's chapter examines the environmental problems that have accompanied South Korea's economic growth and the role of the government in dealing with these problems. Shin notes that until recently government officials and many Koreans in general viewed pollution simply as an unavoidable by-product of economic growth. Changes in the government's and the public's attitude have come about only within the last few years and the central government has taken some steps to counter environmental degradation. However, questions remain about the extent to which government actions will be sufficient. Moreover, as Shin notes, with greater devolution of political power, the role of grass-roots environmental groups will be more important than ever.

The final chapter, by Suntaree Komin, surveys Thailand's considerable environmental problems and explores people's attitudes in Thailand towards the environment and how these attitudes shape decisions concerning the environment. Komin points out that while the religious world view of Buddhism, to which most Thais nominally adhere, teaches living in harmonious alignment with nature, this has not stopped Thais in all walks of life from severely degrading their environment. Komin finds that the attitudes toward the environment of those surveyed are shaped more by self-centered (or, at best, family-centered) notions of materialism and commercialization, coupled with the view that waste is gone once it is off of one's property. The situation is made even worse by lax enforcement of environmental laws. To end on an optimistic note, Komin believes that improvements are possible with greater democracy and through the efforts of those who are concerned about the environment to draw public attention to problems and possible solutions.

Notes

1. Under the category of Asia I am including the regions generally referred to as East Asia, Southeast Asia, South Asia, as well as the Southwest Pacific.

2. Samuel Brittan, writing in the *Financial Times* (11 August, 1990), is critical of such prophets of doom, arguing that eco-doom is based on belief in a form of bogus long-termism: "The malady consists of diagnosing false scares and threats to mankind, based on fashionable preoccupations dressed up as scientific predictions." By way of earlier examples, he cites the work of British economist Stanley Jevons in the mid-nineteenth century on the depletion of fossils fuels and that of Thomas Malthus on the population threat which Malthus linked to a call for "moral restraint." Brittan is skeptical of global environmental threats, but he recognizes that it would be wrong to ignore them, while commenting that actual problems and solutions "are less exciting than eco-doom."

3. Robert Nadelson, "China: The ruined earth," in *Far Eastern Economic Review*, 19 September, 1991, p. 39.

4. There are numerous critical studies of deforestation and forest management in Asia. Among these are: James Rush, *The Last Tree: Reclaiming the Environment in*

Tropical Asia (New York: The Asia Society/Boulder: Westview Press, 1991); Philip Hurst, *Rainforest Politics: Ecological Destruction in South-East Asia* (London: Zed Books, 1990); *The Battle for Sarawak's Forest* (Penang: World Rainforest Movement and Sahabat Alam Malaysia, 1989); François Nectoux and Yoichi Kuroda, *Timber from the South Seas: An Analysis of Japan's Tropical Timber Trade and Local Transformation* (Washington, DC: World Wildlife Fund International, 1989); Mark Poffenberger, ed., *Keepers of the Forest: Land Management Alternatives in Southeast Asia* (Manila: Ateneo de Manila University Press, 1990); M.V. Nadkarni, *The Political Economy of Forest Use and Management* (New Delhi: Sage Publications, 1989); Ramachandra Guha, *The Unquiet Woods: Ecological Change and Peasant Resistance in the Himalaya* (New Delhi: Oxford University Press, 1989); Vandana Shiva, *Forestry Crisis and Forestry Myths: A Critical Review of Tropical Forests: A Call for Action* (Penang: World Rainforest Movement, 1987); and Robert S. Anderson and Walter Huber, *The Hour of the Fox: Tropical Forests, the World Bank, and Indigenous People in Central India* (Seattle: University of Washington Press, 1988).

5. See, Marites Danguilan-Vitug, "Fighting for life," in *Far Eastern Economic Review*, 13 June, 1991, pp. 52-53; and Francisco A. Magno, "Forests, community and development: The political economy of resource use in the Philippine uplands," paper presented to the Fifth Annual Conference of the Northwest Regional Consortium for Southeast Asian Studies, Vancouver, 16-18 October, 1992.

6. According to the World Resources Institute (cited in *Far Eastern Economic Review*, 6 February, 1992, p. 14), the rate of deforestation in Southeast Asia per year are: Indonesia 9,000 square kilometers, Burma 6,770 square kilometers, Malaysia 2,550 square kilometers, Vietnam 1,730 square kilometers, Thailand 1,580 square kilometers, the Philippines 1,430 square kilometers, Laos 1,000 square kilometers, and Cambodia 250 square kilometers (excludes replanted areas).

7. The statement is from the Burma Action Group, quoted in Dhira O, "Relying on resources," in *The Nation* (Bangkok), 1 February, 1993, pp. C1-C2. The Burmese regime focused on the sale of concessions for exploitation of oil and gas, timber, minerals, and fish, and spent most of the foreign exchange earned on the military. Oil and gas exploration accounted for sixty-five percent of foreign investment. Officially recorded gold production increased from 4,000 ounces in 1990-91 to 49,200 ounces in 1992.

8. Dhira O, "Relying on resources." The United Nations estimated the rate of deforestation at 1.2 million acres per year, while World Watch placed the rate at two million acres.

9. Apisak Dhanassettakorn, "Thai loggers in Burma forming body in bid for concessions, in *The Nation*, 25 March, 1993, p. B2; and "New timber body formed," in *Bangkok Post*, 24 March, 1993, p. 15.

10. United States forces spread 11.2 million gallons of Agent Orange (which contains carcinogenic dioxin) and eight million gallons of other defoliants; Liane Clorfeno Caston, "The dioxin file: Anatomy of a cover-up," in *The Nation* (New York), 30 November, 1992, pp. 658-662, 664.

11. On the logging ban in Laos, see, "Laos bans logging pending new controls," in *Far Eastern Economic Review*, 3 October, 1991, p. 59. Logging in recent years has been the country's largest export earner (US$45 million in 1988, representing

about half of total export earnings). The ban was imposed to allow the government to take steps to avoid destruction of the resource.

12. On logging in Burma, see, Bertil Lintner, "Burmese plumder," in *Far Eastern Economic Review*, 4 June, 1992, p. 63.

13. See Ken Stien, "Log rolling," in *Far Eastern Economic Review*, 21 January, 1993, pp. 15-16. The United Nations imposed sanctions on the export of timber and mineral products from Khmer Rouge territory in early 1993.

14. Victor Mallet, "Cambodia powerless to prevent plundering of its forests," in *Financial Times*, 20 August, 1992, p. 4. Also see, Jon Liden and Murray Hiebert, "Cambodian assault," in *Far Eastern Economic Review*, 4 June, 1992, p. 64; and "Tree thieves," in *Far Eastern Economic Review*, 11 June, 1992, p. 7. Regarding Burma, see, "Border talks," in *Far Eastern Economic Review*, 8 October, 1992, p. 9.

15. The activities of Indonesia's leading forestry enterprise, Barito Pacific Group, and its president, Prajogo Pangestu, is discussed in a series of articles in the *Far Eastern Economic Review*, 12 March, 1992, pp. 42-46. Plans by Scott Paper and PT Astra International to develop a large pulp and paper project in 1988 was the subject of an international campaign to have the company halt the project. Scott Paper withdrew from the project in 1989 and Astra withdrew in 1992 (see, "Astra withdraws from paper project in Irian Jaya," in *Jakarta Post*, 10 August, 1992, p. 1).

16. Anthony Rowley, "Ravaged continent," in *Far Easern Economic Review*, 21 November, 1991, p. 66.

17. Martin W. Lewis, *Wagering the Land: Ritual, Capital, and Environmental Degradation in the Cordillera of Northern Luzon, 1900-1986* (Berkeley: University of California Press, 1992), pp. 3-4. Also see, H. Ramachandran, ed., *Environmental Issues in Agricultural Development* (New Delhi: Concept Publishing Company, 1991).

18. See, for example, Piers Blackie and Harold Blakefield, eds., *Land Degradation and Society* (London: Metheun, 1987), pp. 37-48.

19. See, for example, J. Peter Brosius, *After Duwagan: Deforestation, Succession, and Adaptation in Upland Luzon, Philippines*, Michigan Studies of South and Southeast Asia No. 2 (Ann Arbor, MI: Center for South and Southeast Asian Studies, University of Michigan, 1990).

20. See, Martin, *Wagering the Land*, pp. 184-186.

21. Victor Mallet, ""Poisoned chalice," in *Financial Times*, 24 March, 1993, p. 13, discusses the controversy over the provision of such pesticide to Cambodia by Japanese aid agencies. The Japanese have been accused of providing the pesticides to help their chemical industry secure a foothold in this new market.

22. See, Michael C. Howard, *Mining, Politics, and Development in the South Pacific* (Boulder, CO: Westview Press, 1991), pp. 64-65.

23. See, Howard, *Mining, Politics, and Development in the South Pacific*, chapter 6; Christopher Weeramantry, *Nauru: Environmental Damage Under International Trusteeship* (Melbourne: Oxford University Press, 1992); Ian Williams, "Where the Phosphate and Time are Running Out," in *Pacific Islands Monthly*, March, 1991, pp. 19-22; and "Nauru's experience after strip-mining," in *Pacific Islands Monthly*, July, 1992, p. 27.

24. "Quenching the tigers' thirst," in *The Economist*, 15 August, 1991, p. 21; Anthony Rowley, "Heart of darkness," in *Far Eastern Economic Review*, 28 January, 1993, pp. 44-46.

25. See Carl Goldstein, "China's generation gap," in *Far Eastern Economic Review*, 11 June, 1992, pp. 45-47; Sheryl WuDunn, "Difficult algebra for China: Coal = growth = pollution," in *New York Times*, 25 May, 1992, pp. 1, 5; and "China: Powerless growth," in *The Economist*, 28 November, 1992, pp. 32-33.

26. See, Jose Galang, "Power cuts hit Manila growth aims," in *Financial Times*, 17 June, 1992, p. 4; and Jeremy Clift, "Chronic power shortage affects future growth," in *Manila Bulletin*, 22 June, 1992, p. B1; Rigoberto Tiglao, "Bent over backwards," in *Far Eastern Economic Review*, 28 January, 1993, p. 47; Jose Galang, "Ramos takes emergency powers to cope with crippling power cuts," in *Financial Times*, 6 April, 1993, p. 6; and Jose Galang, "Philippines financing breaks new ground," in *Financial Times*, 30 April, 1993, p. 5.

27. See "Brown and out in Manila," in *The Economist*, 10 April, 1993, pp. 25-26; and Jose Galag, "Manila set to deal with crippling power shortages," in *Financial Times*, 30 March, 1993, p. 6. Also see, Rigoberto Tiglao, "Days of darkness," in *Far Eastern Economic Review*, 24 October 1991, pp. 68-69.

28. See, for example, Michael C. Howard, *The Impact of the International Mining Industry on Indigenous Peoples* (Sydney: Transnational Corporations Research Project, University of Sydney, 1988), p. 65, 235; "Existing and proposed hydro-electric dam projects affecting Philippine minorities," in *Sandugo*, 3rd/4th Quarters, 1983, pp. 28-29; "Philippines tribal minorities threatened by Chico dams project," in *Survival International Review*, Vol. 5, Nos. 3/4, 1980, pp. 39-41.

29. See, Doug Tsuruoka, "Rumble in the jungle," in *Far Eastern Economic Review*, 18 June, 1992, pp. 74-75; Doug Tsuruoka, "Cross currents," in *Far Eastern Economic Review*, 30 July, 1992, p. 53; and Kieran Cooke, "Malaysia dam dispute opens floodgates of hostility," in *Financial Times*, 21 July, 1992, p. 5.

30. See, Paul Handley, "Power struggles," in *Far Eastern Economic Review*, 17 October, 1991, pp. 98-99. Public protests in February and March 1993 generated considerable controversy concerning the dam. The events are covered in the *Bangkok Post* and *The Nation*, and, especially, the 21 March, 1993, issue of the *Sunday Post* (pp. 17, 20), that includes a chronology of events and articles examining both sides of the debate.

31. Yvonne Preston, "A million must move for China's next Great Wall," in *Financial Times*, 7 April, 1992, p. 4.

32. See, B.D. Dhawan, ed., *The Dig Dams: Claims and Counter Claims* (New Delhi: Commonwealth Publishers, 1991); *Before the Deluge: Human Rights Abuses of India's Narmada Dam* (Washington, DC: Asia Watch, 1992); Stephanie Gray, "World Bank admits Indian dam flawed," in *Financial Times*, 20/21 June, 1992, p. 6; Susumu Awanohara and Rita Manchanda, "Dam under fire: World Bank report rekindles environmental row," in *Far Eastern Economic Review*, 2 July, 1992, p. 18; "Health impact ignored," in *Economic and Political Weekly*, 18 July, 1992, pp. 1513-1514; Stefan Wagstyl, "India's dam-busters," in *Financial Times*, 30 September, 1992, p. 14; George Graham, "World Bank to decide fate of Narmada Dam," in *Financial Times*, 22 October, 1992, p. 6; Stefan Wagstyl, "World Bank ready to drop $3 bn dam," in *Financial Times*, 30 March, 1993, p. 6; and Hamish McDonald,

"Closing the flood gates," in *Far Eastern Economic Review*, 15 April, 1993, p. 15. Also see a full-page advertisement placed by opponents of the project (the advertisement lists 250 groups from a variety of countries) under the title "The World Bank must withdraw immediately from Sardar Sarovar," in *Financial Times*, 21 September, 1992, p. 5.

33. A recent survey of nuclear power in Asia is provided by Frank Gray, "Ambitious schemes," in *Financial Times*, 15 October, 1992, pp. 14-15. Also see, Doug Tsuruoka, Suhaini Aznam, and Carl Goldstein, "Power plays," in *Far Eastern Economic Review*, 28 January, 1993, pp. 48-49.

34. Carl Goldstein, "The nuclear option: China's programme dogged by delays," in *Far Eastern Economic Review*, 11 June, 1992, pp. 50-51.

35. See, "Indonesia looks to nuclear power," in *Sunday Observer* (Jakarta), 16 August, 1992, p. 1. On Indonesian power requirements in general, see, William Keeling, "Private surge in Indonesian power sector," in *Financial Times*, 9 July, 1992, p. 7.

36. See "Panel looks into nuke plant," in *Bangkok Post*, 10 February, 1993, p. 8. The chair of the House Energy Committee is quoted in the article as calling for a feasibility study since "bunker oil and lignite fuels had polluted the environment and there were restrictions on building hydroelectricity dams."

37. On urban growth in Asia, see, R.J. Fuchs, G.W. Jones, and E.M. Pernia, eds., *Urbanization and Urban Policies in Pacific Asia* (Boulder: Westview Press, 1987); N. Ginsburg, B. Koppel, and T. McGee, eds., *The Extended Metropolis: Settlement Transition in Asia* (Honolulu: University of Hawaii Press, 1990); and Terry G. McGee and Charles Greenberg, "The emergence of extended metropolitan regions in ASEAN 1960-1980: An exploratory outline," in A. Pongsapich, M.C. Howard, and J. Amyot, eds., *Regional Development and Change in Southeast Asia in the 1990s* (Bangkok: Chulalongkorn University Social Research Institute, 1992), pp. 133-161.

38. See, Brendan F.D. Barrett and Riki Therivel, *Environmental Policy and Impact Assessment in Japan* (London: Routledge, 1992).

39. *Bangkok Post*, 28 January, 1993, p. 9.

40. Dhira O, "Out of the wasteland," in *The Nation* (Bangkok), 25 March, 1993, pp. C1, C3.

41. See chapter by Suntaree Komin in this volume and Dhira Phantumvanit and Liengeharensit, "Coming to terms with Bangkok's Environmental Problems," in *Environment and Urbanization*, Vol. 1, No. 1, 1989, pp. 31-39.

42. See, Michael Vatikiotis, "Official fog," in *Far Eastern Economic Review*, 14 November, 1991, p. 24.

43. See, for example, Stuart Chape and Dick Watling, *Environment: Fiji: The National State of Environment Report* (Suva: National Environment Management Project, 1992).

44. *Development and the Environment* (Washington, DC: The World Bank, 1992).

45. Victor Mallet, "Stagnant rivers and poisoned seas," in *Financial Times*, 31 March, 1993, p. 14.

46. Chandan Datta, "Yamuna River turned sewer," in *Economic and Political Weekly*, 5-12 December, 1992, pp. 2633-2636.

47. International Development Research Centre, *Aquaculture Economics Research in Asia* (Proceedings of a Workshop Held in Singapore, 2-5 June 1981]; T.E. Auon and D. Pauly, eds., *Coastal Area Management in Southeast Asia: Management Strategies and Case Studies: ICLARM Conference Proceedings* (Kuala Lumpur: Ministry of Science, Technology and the Environment, 1989); *Coastal Management in Pak Phanang: A Historical Perspective of the Resources and Issues* (Hat Yai: Coastal Resources Institute, Prince of Songkla University, 1991); Erik Davies, Sustainable Development Evaluation: The Case Study of Pond Aquaculture in South Sulawesi, Indonesia (M.A. thesis, University of Guelph, School of Rural Planning and Development, 1990).

48. *Coastal Management in Pak Phanang*, p. 64.

49. See, P. Kunstadler and C.F. Bird. *Man in the Mangroves: The Socio-economic Situation of Human Settlements in Mangrove Forests* (Tokyo: United Nations University, 1986). Also see, Padma Narsey Lal, *Conservation of Mangroves in Fiji*, Occasional Paper No. 11 (Honolulu: East-West Environment and Policy Institute, 1990), for a study of mangroves in the South Pacific.

50. "Official statistics culled from various forestry-related sources show that mangrove forests have dwindled to 168,000 from a heft half a million hectares or so in the 1920's" ("DENR starts mangrove rehabilitation," new release from the Public Affairs Office of the Department of the Environment and Natural Resources, 1991).

51. Marsudi Triatmodjo, "Protection of the marine environment against pollutants from pond," unpublished paper, Gaja Mada University. See, "Expert blames offshore drilling for sea pollution," in *Indonesian Observer*, 1 August 1991, p. 4.

52. The figures are as of 1988 based on World Resources Institute data; see *Far Eastern Economic Review*, 30 January, 1992, p. 12.

53. See, Hilary de Boer, "Sticking a neck out to save the crane," in *Financial Times*, 8 April, 1992, p. 12, on competition for land between farmers and wildlife conservationists in Vietnam; and "Conservation at human cost: Case of Rajaji National Park," in *Economic and Political Weekly*, 1-8 August, 1992, pp. 1647-1650, for a case study from India.

54. Philip Dearden, "Tourism and development in Southeast Asia: Some challenges for the future," in A. Pongsapich, M. Howard, and J. Amyot, *Regional Development and Change in Southeast Asia in the 1990s*, p. 215. See, Gregory Hodgson and John A. Dixon, *Logging Versus Fisheries and Tourism in Palawan*, Occasional Paper No. 7 (Honolulu: East-West Environment and Policy Institute, 1988).

55. Dearden, "Tourism and development in Southeast Asia," p. 219. Also see, David Jolly, "Greener beaches: Thailand aims to clean up tourist resorts," in *Far Eastern Economic Review*, 13 August, 1992, p. 58.

56. See, "Berjaya faces Environmental challenge," in *Asian Wall Street Journal*, 17 June, 1992, p. 3.

57. Peter Knight, "Rough time for birdies," in *Financial Times*, 19 August, 1992, p. 10. For criticism of plans to build a golf course near Manila, see a letter by A.R.T. Kemasang, "Hidden costs of golf courses," in *Far Eastern Economic Review*, 6 February, 1992, p. 6. In relation to Thailand, see, A. Plaumanom, "Golfers

dream, farmers nightmare," in *Regarding Tourist Development in Thailand*, special issue of *Thai Development Newsletter*, No. 20, 1991-1992, p. 34.

58. Murray Hiebert, "Green fees: Environmentalists oppose a new golf course," in *Far Eastern Economic Review*, 20 August, 1992, p. 22; Jatuphol Rakthammachat, "Tourism: Saviour or spoiler?, in *The Nation* (Bangkok), 31 January, 1993, pp. B1, B3. In response to protests, logging was suspended on 30 December 1992, but indications were that the suspension was temporary.

59. *Potential Socio-Economic Effects of Climate Change* (Oxford: Environmental Change Unit, Oxford University, 1991). Also see, John Hunt, "Cold comfort on global warming," in *Financial Times*, 8 November, 1991, p. 28; and Paul Handley, "Before the flood," in *Far Eastern Economic Review*, 16 April, 1992, p. 65-66.

60. See, Robert Keith-Reid, "Forum 19: New name and a new role," in *Islands Business*, October, 1988, p. 14; Robert Keith-Reid, "Forum opens fire in the second battle of Tarawa," *Islands Business*, August, 1989, p. 18; Peter Roy and John Connell, The greenhouse effect: Where have all the islands gone?," in *Pacific Islands Monthly*, April/May, 1989, pp. 16-21; and Ian Williams, "Success for Pacific nations at Rio," in *Pacific Islands Monthly*, July, 1992, p. 25.] The South Pacific Forum has also expressed concern with plans by the United States to incinerate nerve gases on Johnston Island, drift-net fishing, and French nuclear testing in French Polynesia.

61. Source of figures (which are for 1988), World Resources Institute, *World Resources Report 1990-91* (New York: Oxford University Press, 1990); also see, *The Economic Costs of Reducing CO2 Emissions*, OECD Economic Studies No. 19 (Paris: OECD, 1992).

62. See footnote 18 in the chapter by Howard on mining in this volume concerning the establishment of regulatory agencies in the mining industry in Asia. For discussions of the establishment of forestry agencies in Asia, see Poffenberger, *Keepers of the Forest*, on Southeast Asia; and Nadkarni, *The Political Economy of Forest Use and Management*, on India.

63. Husin Ali, "Non-governmental organizations," in Michael C. Howard and Ted Wheelwright, eds., *The Struggle for Development: Essays in Honour of Ernst Utrecht* (Burnaby, BC: International Studies Programme, Simon Fraser University, 1990), p. 77.

64. See the chapter by Devalle in this volume; Guha, *The Unquiet Woods*; and Rush, *The Last Tree*, pp. 56-60.

65. The Sahabat Alam Malaysia (Friends of the Earth Malaysia) was initially a part of the Consumers' Association of Penang which was founded in 1969, and it became a separate entity in 1977. The Environmental Protection Society Malaysia was founded in 1974. There is also the Malayan Nature History Society, which was founded in 1940, but did not become a vocal critic of environmentally destructive acts until the 1970s. See, Rush, *The Last Tree*, pp. 69-73.

66. Barrett and Therivel, *Environmental Policy and Impact Assessment in Japan*.

67. Handley and Awanohara, "Thai dam scheme causes split."

68. See, Ali, "Non-governmental organizations," p.83. Antonio L. Ledesma ("Nongovernmental organizations and the limits of state power, in *Solidaridad*, No. 127, July-September, 1990, p. 61) states that there were 16,000 nongovernmental

organizations in the Philippines in 1985 and that another 5,000 were formed during the next three years.

69. "Environmentalists have no right to cause trouble: M'sia official," in *Indonesian Observer*, 8 July, 1991, p. 6.

70. See, Geoffrey Pleydell, "'Greens' given hard time at rainforest conference," in *Financial Times*, 16 October, 1992, p. 34.

71. "RI denies reports on destruction of tropical forests," in *Jakarta Post*, 8 August, 1992, p. 8.

72. Quoted in Adam Schwarz, "Timber is the test," in *Far Eastern Economic Review*, 23 July, 1992, p. 36. Also see, Michael Vatikiotis, "Malaysia's war," in *Far Eastern Economic Review*, 4 June, 1992, p. 65.

73. "Scott paper project recycled," in *Tok Blong SPPF*, No. 30, 1990, p. 19.

74. This method is being employed in a joint venture between Bureau de Recherches Géologiques et Minières of France and Barclays Metals of the United Kingdom at the Kilembe copper mine on the edge of Queen Elizabeth National Park in Uganda to extract cobalt (Kenneth Gooding, "Uganda's biological cobalt project," in *Financial Times*, 29 January, 1992, pg. 22).

75. J.P. Cooney, "A business perspective on trade and environment," unpublished speech to Globe '92 panel of Trade and Environment, Vancouver, 19 March, 1992.

76. In Indonesia, BP Solar, with World Bank backing, is involved in efforts to spread the use of solar power generation, especially to more isolated areas; see, "Power to the people," in *Far Eastern Economic Review*, 22 April, 1993, p. 56; and William Keeling, "Throwing light on Indonesia," in *Financial Times*, 25 March, 1993.

77. See, Joe Kirwin, "Hot air fuels the energy debate," in *Financial Times*, 19 May, 1993, p. 16.

78. See, *Coal-use Technology in a Changing Environment*, Financial Times Management Report (London: Financial Times, 1992).

79. In Thailand, electricity authorities launched a major drive to promote more efficient energy use in early 1993; see, "Drive to save Thailand Bt60 bn by year 2001," in *The Nation* (Bangkok), 25 March, 1993, p. B3.

80. Ian Rodger, "Cleaner dyes make Jakarta less colourful," in *Financial Times*, 31 October, 1991, p. 10.

81. There are also questions concerning just how far technology alone within a single industry can solve environmental problems. For example, in response to efforts by the auto industry to reduce pollution through the use of catalytic converters, recycling of components, and meeting more exacting exhaust emission standards, lobbyist Stephen Joseph argues that the limits of such efforts may have been reached and that an option such as the use of electric cars merely shifts "from street to power station" ("The climate is changing," in *Financial Times*, 20 October, 1992, p. XI).

82. See the chapter by Kumar. Also see, Vandana Shiva, "Biotechnology development and conservation of biodiversity, in *Economic and Political Weekly*, Vol. 26, No. 48, 1991, pp. 2740-2746, which refers to "colonisation of the seed" in relation to the development of high-yielding varieties and biodiversity.

83. See, George Wehrfritz, "Taiwan" Asia's richest, but also dirtiest," in *Far Eastern Economic Review*, 29 October, 1992, p. 38.

84. Jamie Allen, "Hongkong: Set in a septic sea," in *Far Eastern Economic Review*, 19 September, 1991, p. 40.

85. See, Kieran Cooke and Angus Foster, "UK water group in Malaysia deal," in *Financial Times*, 20 April, 1993, p. 19; Doug Tsuruoka, "Where there's muck," in *Far Eastern Economic Review*, 1 April, 1993, p. 79; Mark Clifford, "Making polluters pay," in *Far Eastern Economic Review*, 14 January, 1993, p. 43; William Keeling, "Indonesian approval for waste plant project," in *Financial Times*, 5 February, 1993, p. 5; Jearanai Thasai, "KC Group gets on environmental wagon," in *The Nation* (Bangkok), 9 April, 1993, p. B11; and Angus Foster, "UK company secures Bangkok sewage deal," in *Financial Times*, 1 April, 1993, p. 5.

86. On the history of government environmental management in Thailand see, Sriracha Charoenpanij, "The Thai legal system: The law as an agent of environmental protection," in *Culture and Environment in Thailand: A Symposium of the Siam Society* (Bangkok: The Siam Society, 1989), pp. 463-473; and Anat Arbhabhirama, et al., *Thailand Natural Resources Profile* (Singapore: Oxford University Press, 1988).

87. Paul Handley, "Thailand: New rules, but old attitudes," *Far Eastern Economic Review*, 29 October, 1992, p. 40.

88. See, Duangkamol Chotana and Thitinan Pongsodhirak, "TDRI's Anand takes in top post at green think tank," in *The Nation* (Bangkok), 24 March, 1993, p. B1.

89. *Far Eastern Economic Review*, 7 November, 1992, p. 8.

90. Nadelson, "China: The ruined earth."

91. See, Poffenberger, *Keepers of the Forest*. Also see, Michael Redclift, "A framework for improving environmental management: Beyond the market mechanism," in *World Development*, Vol. 20, No. 2, 1992, p. 256.

92. Anan Ganjanapan, "Community forestry management in northern Thailand," in Pongsapich, Howard, and Amyot, *Regional Development and Change in Southeast Asia in the 1990s*, p. 76.

93. "Forestry set to give public a greater say," in *Bangkok Post*, 28 January, 1993, p. 2.

94. The failure of governments in the Asia-Pacific region to find a satisfactory means of promoting participatory development that is sensitive to ethnic minorities is discussed in Michael C. Howard, "Ethnicity, development, and the state in Southeast Asia and the Pacific," in Pongsapich, Howard, and Amyot, *Regional Development and Change in Southeast Asia in the 1990s*, pp. 91-104.

95. M. Shugenne, "Crisis in theory," in *Economic and Political Weekly*, 18 May, 1991, p. 1274.

96. Jefferson Plantilla, "Community forest management: Toward strengthening the community stewardship agreement," in J. Fox, O. Lynch, M. Zimsky, and E. Moore, eds., *Voices from the Field: Fourth Annual Social Forestry Writing Workshop* (Honolulu: East-West Center, 1991), p. 75.

97. John McBeth, "Forest of family trees," in *Far Eastern Economic Review*, 27 September, 1990, pp. 36-37. A multi-sectoral committee was established in an effort to resolve the differences.

98. See the chapter on small-scale mining by Howard in this volume. The cases of Nauru and its phosphate mining is an interesting example of a local population taking over a large mining enterprise, but one must be cautious in generalizing from this rather unusual case, while also noting that it is far from being an unequivocal success story (see, Howard, *Mining, Politics, and Development in the South Pacific,* chapter 6).

2

Colonization of Tribal Lands in Bangladesh and Indonesia: State Rationales, Rights to Land, and Environmental Justice[1]

G. Peter Penz

Roughly two decades ago, two independent political events allowed the states of Bangladesh and Indonesia to take control of two regions that have been occupied by tribal groups traditionally engaged in subsistence production.[2] These regions are the Chittagong Hill Tracts and western New Guinea, the latter referred to as Irian Jaya by Indonesian authorities, but as West Papua by indigenous nationalists. The events were the secession of Bangladesh from Pakistan and the incorporation of Dutch West New Guinea by Indonesia as Irian Jaya. Both became instances of state-sponsored colonization of tribal areas by settlers from the overpopulated heartland of these two countries. Both are now reaching or passing the point where the indigenous population is becoming the minority. They have been described as "two of the most destructive *development by invasion* transmigration programs" being conducted.[3] This chapter, after describing these cases, addresses such a state policy through ethical analysis, posing the question of whether the colonization of tribal areas and the accompanying exploitation of their resources can be justified.

The issue that is being addressed here concerns the responsibilities of the state regarding the use of land and rights to it. Justifications offered by states engaged in this process of development by colonization will be assessed and that will be done from the perspective of environmental justice. However, environmental justice is to be understood in the broadest sense of that term.[4] The focus is on human interests in and claims to the environment, and particularly land; the intrinsic value of the environment is considered only briefly.

I will proceed as follows. After a thumbnail sketch of the situations in the Chittagong Hill Tracts and Indonesian Papua and the identification of the paradigmatic features of this kind of situation, I will defend ethical analysis against certain criticisms, including that delegitimation of state action through description is sufficient, and indicate the general approach to be adopted, which is one of practical policy ethics. This brings the chapter to its central focus, which is the ethical evaluation of state rationales for the colonization and resource exploitation of tribal frontier lands. Accordingly, I present a classification of policy rationales, and this provides the organization for the remainder of the chapter. The rationales are in terms of (1) national development, (2) environmental protection, (3) alleviating heartland poverty, (4) developing "backward" peoples, and (5) territorial integrity and national unity. The conclusion will be that none of these policy rationales survives close scrutiny, that the colonization of tribal frontier lands and the exploitation of their resources is not justified and that, moreover, given what has happened to them, the tribal peoples have the moral right to secede.

Colonizing Tribal Lands in Bangladesh and Indonesia

The Chittagong Hill Tracts

The Chittagong Hill Tracts are located in the southeast corner of Bangladesh, away from the coast, and stretch from the south along the border with Burma and India's state of Mizoram for about 250 kilometers into a cul-de-sac formed by the Indian state of Tripura. It constitutes ten percent of the total area of Bangladesh and holds potentially large reserves of mineral gas and petroleum and also coal and copper. There are about 600,000 hill people living in a terrain of forested hills and valleys. They constitute less than one percent of the population of Bangladesh, the rest of which is nearly totally Bengali and is eighty-five percent Muslim. Bangladesh is thus one of the most homogeneous nation states in the world. The relatively tiny minority of hill peoples, however, is culturally and racially quite distinct from the Bengalis. The hill tribes are of Sino-Tibetan descent, like the hill people on the other side of the border, and are mostly Buddhist, with the remainder Hindu, Christian, and animist, rather than Muslim. They consist of thirteen tribes, but the Chakma, the Marma, and the Tippera make up nearly ninety percent of this population. They have been relying traditionally on shifting cultivation, with the land held in common and only usufruct rights for families, although now wet rice cultivation is practiced quite extensively in those flat areas suitable for it.[5]

In the past, it seems that they were entirely independent politically, until the region was annexed by the Mughal regime of India, but only to levy a trade tax. Following its penetration by the British in the 1860s the restrictions on the region's political self-determination were fairly limited, although the practices of slavery, debt bondage, headhunting, inter-tribal warfare, and raids on the coastal people were banned. After initially encouraging Bengali immigration, the British changed their policy when the Indian independence movement took root among Bengalis and formalized tribal autonomy with the Chittagong Hill Tracts Regulation of 1900, which strictly controlled the entry and residence of non-tribals. In 1935 the British declared the region a "totally excluded area," with its administration reporting to the central government of British India rather than to the provincial government of Bengal. Tax-collecting, policing, and many aspects of criminal and civil law were administered by the tribal chiefs. One important intervention that the British did make was in the 1870s, when they banned the then ubiquitous shifting agriculture from about one-quarter of the forest land and limited the land a tribal family could hold. This, together with tribal population growth, led to more intensive cultivation of the land, the halving of the fifteen-to-twenty-year fallow period in the cultivation cycle, and some deterioration of soil in the hills as well as the emergence of settled agriculture in suitable areas. The British also opened the hills up to commercial activities. At the time of the independence negotiations for India, however, there was a movement for either an independent state, or inclusion in the Indian Union, but in the end the Chittagong Hill Tracts were "awarded" by the British to Muslim Pakistan and incorporated into what then became East Pakistan.

Under the Pakistani state, the process of the foreign-aid-assisted "development" of the Hill Tracts began. The indigenous tribal police force was disbanded. Twenty-two percent of the Hill Tracts were set aside as a forest "reserve" for tropical-hardwood lumbering and for pulp bamboo for a new paper mill in the region. When the East Pakistani government, the junior government in the federal system of Pakistan, tried to fully incorporate the Hill Tracts into its administrative system in 1955, resistance from local administrators and inhabitants led to greater control from the central government in Karachi and reinstitution of the Hill Tracts' status as an "excluded area." With the military take-over in Karachi in 1958, the interest in opening up the region gained new momentum. In the early 1960s the United States-supported and -constructed Kaptai Dam, a hydroelectric project, was completed and the reservoir filled, submerging 650–700 square kilometers of prime agricultural land, about forty percent of the cultivable land of the Hill Tracts, and displacing over 100,000 of the indigenous population. Only very partial compensation in land was provided to two-thirds of this group, the remainder being forced onto the already overpopulated hill slopes, with a resulting impairment of soil fertility. A survey in 1979 indicated that ninety-three percent of the tribespeople considered themselves to be worse off than before the dam was built. In 1964, two years after protection for the area had been given constitutional status, the government abolished the Hill Tracts' special status with its tribal land rights and immigration restrictions. Indigenous officials and police were replaced by non-local agents of law and order. A steady stream of poor Bengali settlers into the district began, but it was on a relatively small scale. An international team of consultants was enlisted to devise a master plan for integrated development of the region with the aim of optimum land use. The initiation of a scheme to transform the Hill Tracts into fruit gardens and softwood plantations and to reduce shifting cultivation led to local famines and cases of starvation.[6] This period has been described as one of "tension between efforts to preserve the traditional way of life found in the Tracts, and on the other hand efforts to open the area to migration and economic development by the Bengali."[7]

The situation became critical with the Bangladesh war of liberation from Pakistan. Minor hill-tribe groups fought on both sides and several tribal chiefs cooperated with the Pakistanis to protect their positions as chiefs or their people from being victimized by the Pakistani army. Although the majority tried to stay out of the conflict, the Bengali perception was one of hill-tribe support for Pakistan. As some hillpeople fled to India to escape the war or punitive raids by Bengali freedom fighters and the Bangladesh army, 30,000 to 50,000 Bengalis entered the northern subdivision of the Chittagong Hill Tracts and occupied their lands. When the hill-tribe leader, Manabendra Larma, led a delegation to meet the new prime minister, Sheik Mujibur Rahman, they were told to discard their ethnic identities and become Bengalis. A policy of offering land in the Hill Tracts to Bengalis and settling them there was initiated. In 1979 President Ziaur Rahman put in motion a plan to settle Bengalis in the region on a massive scale. The government considered much of the land to be state property to be distributed at the discretion of state officials. Land was allocated to each settler family and rations and grants, partly financed by foreign aid, were provided to help them through their first year.[8] No compensation was given to the hill tribes. By 1982, between 300,000 and 400,000 Bengalis had been settled in the Hill Tracts. Official estimates indicate that by 1981 the tribals were only about sixty percent of the region's population (compared to ninety-eight percent in 1947), and the time when they (now numbering about 600,000) are turned into a minority is either imminent or has already passed.[9]

This process of Bengali colonization, with the accompanying evictions and atrocities, led to the formation of a hill-tribes guerrilla force, called the Shanti Bahini, in 1975. They have demanded autonomy within Bangladesh, with a separate legislature, a ban on further immigration and the restitution of all lands taken by Bengali settlers since 1970. Their regulars were estimated in 1989 to number about 5,000, apart from local militia.[10] It has been getting limited support from India, apparently in retaliation for alleged Bangladeshi support for the Tripura National Volunteers rebelling against Indian authorities. Founded as a "home guard" against the pressure of Bengali immigrants from the plains, it eventually went on the offensive, kidnapped officials and attacked Bengali settlements, disrupted communications, and even resorted to assassinating fellow tribals who collaborated with the authorities.[11] In 1984, for example, the Shanti Bahini killed eighty Bengali settlers and wounded 800 more in an attack on two settlements.[12]

The Bangladeshi government responded to these developments by stationing about 120,000 troops in the Hill Tracts by the early 1980s.[13] Quite apart from direct military assistance from the United States and Britain, foreign development assistance for transport, communication, and relocation projects were used for what the tribals deemed to be military and colonization purposes.[14] Settler aggression in the form of village burnings, mass killings, and related atrocities, often in retaliation for tribal guerrilla actions and frequently with at least lower-level military support, which began in 1977, escalated, with about 300 unarmed tribespeople killed in Kaokhali Bazar in March 1980 and about 500 around Matiranga in June 1981.[15] Religious persecution has taken the form of violence against Buddhist monks and the desecration of temples.[16] The estimate, by the International Work Group for Indigenous Affairs, of the number of tribals killed by 1984 was 185,000 and by 1987 it was 200,000.[17] If these figures are at least indicative of the general magnitude of the killings, they indicate a truly genocidal rampage.[18] Many tribals have been relocated in what some describe as "concentration camps"[19] and others have fled across the border into India, in 1991 their being an estimated 63,000 refugees in the state of Tripura,[20] in spite of very poor conditions in the refugee camps, and an additional unknown number in the state of Mizoram.[21] In the mid-1980s between 5,000 and 10,000 were estimated to be imprisoned, with torture used to obtain information and confessions.[22]

During this escalation, some high government officials favored granting local autonomy, but this was resisted by the parliament. After a military coup brought Ershad to power, the government declared an amnesty for the Shanti Bahini in 1983, with an offer of land, jobs, and cash. It seems that at the time less than one hundred guerrillas surrendered, but two years later a group of 235 turned themselves in, although there is a claim that several thousand rebels surrendered.[23] From 1985 on, there have been contacts from time to time between the Bangladesh authorities and the Shanti Bahini, but with little progress, the demanded expulsion of Bengali settlers being the basic bone of contention. In 1986 the army, which was then said to be "one of the most poorly controlled armies in Asia,"[24] was disciplined, apparently to prevent further reprisals and atrocities. At the same time, the government has made an effort to accelerate development of the region and to assist tribals in adjusting to settled agriculture.[25] It also tried to persuade the refugees in India that it was safe to return and offered assistance, but, it seems, with relatively little success.

Early in 1989 Dhaka approved limited local autonomy in the Chittagong Hill Tracts, with three elected district councils. The measure provides for tribal chairpersons and control over the police, local taxation, and the transfer of land rights, including the rectification of fraudulent or corrupt appropriations.[26] However, development areas are excluded from the councils' jurisdiction and all important decisions have first to be approved by the government of Bangladesh, which can also dissolve the councils.[27] The law was rejected by the Shanti Bahini, which boycotted the elections. In May of 1989 renewed violence broke out when the Village Defence Party, a government-sponsored civilian defence force of settlers trained by the police, attacked six tribal villages and put fire to them. As a result, a new wave of about 6,000 tribal refugees crossed into India.[28] At the same time there were certain hopeful signs. Under the military dictator, General Ershad, an independent commission of experts in international law and tribal communities was given permission to visit the area and it subsequently proposed certain further autonomy provisions.[29] With now a new regime in power in Bangladesh, India has requested renewed efforts to create suitable conditions for the repatriation of the Chakma refugees. To what extent the new democratic government will respond remains to be seen.

"Transmigration" to Indonesian Papua

In Indonesia, state-sponsored colonization of the outer islands of the Indonesian archipelago is called "transmigration." It involves the movement of mostly poor people from overpopulated Java, with its population of over 115 million people (some two-thirds of Indonesia's population on seven percent of its land area) as well as the adjacent smaller islands of Madura and Bali. Some transmigration had occurred under the Dutch colonial administration starting in 1905, but it was very limited in scale.[30] It was only after the military regime replaced the Sukarno government in 1965 that it took a more structured form. One estimate is that in the 1950s and 1960s about 400,000 persons moved under transmigration, 800,000 in the 1970s, and 2.9 million between 1980 and 1985. The transmigration target for the five-year plan for 1984-89 was 750,000 families.[31] This program has been supported by United Nations agencies, the World Bank, and regional development banks as well as several Western nations, with as much as US$600 million committed by the mid-1980s.[32] It is also a strategy by the Indonesian state of political and cultural integration, which leaves no room for regional self-determination and which requires that, in Burger's words, "the distinct peoples of Kalimantan, South Moluccas, East Timor or West Papua must either conform to the Javanese notion of the state or disappear." Transmigration serves to create loyal majorities in areas where the local population resists integration.[33] Nietschmann and Gault-Williams have reported that there is a plan to move an additional sixty-five million people from the central islands to less populated outer islands over the next twenty years.[34]

Indonesian Papua, named Irian Jaya by the government, has, since 1984, become one of the targets for transmigration. It is rich in deposits of oil and of minerals, such as nickel, silver, gold, copper, cobalt, tin, and molybdenum, and has vast timber resources. This area is populated by about one million Papuans. They are Melanesians and, thus, ethnically quite distinct from the Javanese and other peoples living on the Indonesian islands and in religion are animist and Christian rather than Muslim. At least three-quarters of them live in the highlands, practicing shifting cultivation as well as intensive gardening, while the coastal population makes its living partly off fishing and partly off sago plantations. They are organized in terms of small tribal groups that view each other as foreigners and they speak about 260 different languages. Their customs and physical self-presentation are very different from those of the Javanese and the latter "have nothing but contempt for the indigenous population of the island, whom they regard as virtual savages and treat with scorn."[35]

Permanent colonial control by European powers was not established in Western New Guinea until the nineteenth century, when the Dutch built a fort there and made it a colony of theirs. It was administered separately from the Dutch East Indies. Prior to the Second World War, however, the Dutch presence was limited to a few points along the coast. The use of forced labor and taxation led to some resistance, often in the form of millenarian movements. Early this century the Dutch and other foreign interests began to explore for oil there. In the process they introduced an intermediate class of Indonesian, Chinese, and other Asian functionaries, military and police personnel, and traders that absorbed and dealt with any indigenous resistance. During the Second World War, the Japanese occupied the area, and when the Dutch returned after the Japanese defeat, they faced an Indonesian independence movement that laid claim to all

Dutch-administered territories. However, when independence was finally negotiated in 1949, West New Guinea remained in the possession of the Dutch, who favored eventual independence for it. In the meantime the Dutch began to foster a Papuan elite.

Armed clashes between the Indonesians and the Dutch in 1962 led to new negotiations. The Indonesian position was supported not only by multinational corporations, interested in access to the natural resources in the contested territory, but also by anti-colonialist leaders of the Third World. Pressure on the Dutch by the United States government, for which maintaining Indonesia's role in containing communism was obviously more important than violations of the United Nations Charter, led to the New York Agreement, which was ratified by the United Nations General Assembly. It provided for the transfer of the administration of West New Guinea to the United Nations, which in turn handed it over to Indonesian administration in 1963, with a provision for a plebiscite in 1969 to decide the area's political future. The Indonesian administration immediately dissolved the Papuan parliament, banned all political activity, and destroyed Papuan schooltexts and flags. In 1965, the Free Papua Movement, or OPM (Organisasi Papua Merdeka), was formed, leading to the first major revolt, followed by a number of other uprisings.[36] Indonesian military repression ensued. When the plebiscite was due, it was conducted not on a popular basis, but by using 1,025 appointed representatives. Under explicit threats to the liberty and lives of these representatives, and to the West Papuan people as a whole, they unanimously voted for permanent incorporation into Indonesia. Ignoring the reservations expressed by its own observer, the United Nations accepted the result.[37] Western New Guinea thus officially became a province of Indonesia and was given the name Irian Jaya.[38]

While there is unfair treatment of Papuans in oil and mineral extraction, in the form of environmental damage, lack of compensation, and discrimination in hiring,[39] the main threats to the Papuans come from logging, on the one hand, and from agricultural colonization, on the other hand. In the 1970s, vast logging concessions were granted, mainly to members of Indonesia's ruling elite and to foreign companies, so that by 1980 they claimed more than sixty percent of the region's forests.[40] The lands are legally treated as largely unoccupied and unused, communal land rights are ignored, and little or no compensation is paid for appropriations.[41] Virtually forced labor for Papuans is not unknown. Asmat tribespeople have been drafted to log their own trees for timber companies with the threat of being otherwise treated as subversives and have had wages withheld or not paid at all.[42]

Equally serious is the threat from transmigration. The Jakarta government's plan was to move 685,000 Javanese to Irian Jaya in the five-year plan period of 1984-89, with six hectares of land per newcomer. A decline in Indonesia's oil revenues has slowed down the program and its focus has shifted towards recruiting labor for multinational corporations. So far apparently more than half a million Javanese peasants and retired soldiers have been resettled on Papuan land.[43] If trends continue, it is certainly to be expected that not long after the year 2000 the Javanese will become the majority in Irian Jaya. Although, since about 1985, a policy of "parallel development" has been adopted to raise the living standards of indigenous communities to that of transmigrant settlements, it has led to coercive resettlement of tribals in transmigrant settlements. In effect, indigenous people are forced to give up their land for a small corner of a big settlement.[44]

The expropriation of land led to an escalation of violence, some sporadic and some planned guerrilla action. The OPM launched a series of offensives against the Indonesian army and development projects and, in 1984, it briefly captured the provincial capital of Jayapura. Although at its peak the OPM claimed 30,000 members, it is poorly armed. The Indonesian army dealt with such resistance by using a sophisticated system of counter-insurgency strategy, which involved careful control over the population, including special settlements designed for control and indoctrination, and the creation of a "Java Curtain" of settlements of former military personnel and their families along the border with Papua New Guinea, as well as the use of superior weaponry, including helicopter gunships, and commando units with experience from the bloody conquest of East Timor. This militarization led to the bombing and strafing of Papuan villages and to the killing of tens of thousands of tribals. One figure given is of 150,000 deaths of native people between 1963 and 1983.[45] Arrests, torture, and executions have been carried out without any judicial procedure. The army terror led to the development of a string of refugee camps on the other side of the border with Papua New Guinea, at one time with more than 10,000 West Papuans.[46]

The tragedy for the Papuans has been expressed by Julian Burger as follows:

It is possible with legitimacy to talk about genocide elsewhere—the Mayan Indians in Guatemala, the Ache' in Paraguay, the Chakma and other tribal peoples in Bangladesh—but even in the context of such violence the destruction of the West Papuan people has few parallels. More particularly since until recently the Papuans have had relatively little contact with invading powers and are now experiencing a particularly aggressive and racist colonization from Indonesia...The invasions of the Americas and especially Australia are being reborn in West Papua, only this time aided by the United Nations and the general assembly of nation states.[47]

The Paradigm of the Colonization of Tribal Lands

While much of the detail of the above accounts comes from indigenist advocacy literature and official restrictions on access to the regions has put limits on their verification on the basis of a variety of independent sources, the general pattern is clear. It consists of the paradigmatic case of the developmentalist colonization and extractive exploitation of tribal lands due to the pressure of overpopulation, landlessness, and poverty in the national heartland, on the one hand, and to the drive for industrialization and the quest for cheap raw materials, on the other hand. Having been pushed or kept out of the more fertile regions in earlier historical periods, tribal peoples have settled in areas not suited for intensive cultivation, which has also required them to avoid significant population growth, while the intensification of cultivation and population growth went hand in hand in the national heartland. The latter has also generated, or coexisted with, sharp inequalities, which states now find difficult to address and they try to find other ways of responding to the resulting pressures, including the promotion of migration to the frontier areas. Whether the indigenous inhabitants of these frontier areas resist or not, they are treated, if their existence is acknowledged at all, as security risks or threats, or as obstacles to development, or at least as "backward" peoples in need of help through modernization. Most importantly, land is seen as underutilized, and developing it is deemed the only appropriate thing to do. At the same time, the infrastructure for colonization facilitates the extraction of natural resources, which contributes to the economy's balance of payments or to its industrialization, to the public treasury, and to the wealth of influential sections of the national elite.

Critiquing Policies: Moral Evaluation

Having described the situation, I will now go on to evaluate this pattern of development. To some, this may seem superfluous. Political realists simply hold that states are amoral and that there is no point to evaluating policies ethically. This can be taken to apply particularly to relations with unintegrated tribes, which are merely powerless outsiders. Thus, all we can do, it is thought, is offer explanations for questions such as why power vis-à-vis indigenous peoples is exercised primarily at a certain stage in national development or why it is exercised in the manner that it is. This denies any effectivity to morally motivated efforts to improve the world. All I can say in the context of this chapter is that it is not only implausible, but it would deny much of the accepted rationale for democracy.

There is also a kind of Marxist realism which, however, is more complex. It has a particularly serious problem with ethical analysis, since ethics is seen as part of ideology, and ideology is seen as a distortion of reality effected by dominant-class interests structured by the mode of production. This orthodox Marxist position, however, has itself entangled in the paradox of making this claim, but being clearly ethical itself, in its rejection of the subordination of some human beings by others and, more specifically, in its criticism of capitalism, not only for being historically doomed, but also for being destructive of a truly human existence. However, Marxism as such does not commit one to this view. As Steven Lukes has argued, one can escape the paradox by recognizing that the ideology attribution is aimed at particular approaches to morality and that Marxism does offer "a consistent and distinctive approach to morality and moral questions."[48] More generally, knowledge about social processes, I take it, ultimately serves praxis. But appropriate praxis is determined not merely by how things are or by how structures work; it also depends on notions of how things should be, of what is justifiable. Thus, appropriate praxis requires not only descriptive and explanatory analysis, but also normative or ethical analysis.

Nevertheless, it might still be argued that the immorality of state policy, such as that towards tribal societies, is self-evident, and that all that analysis needs to do is to uncover the detrimental exercise of power that is going on behind the official rhetoric; once the facts are brought into the open, no further analysis, and certainly not ethical analysis, is needed. The history of the colonization of tribal areas, including the recent colonization of the Chittagong Hill Tracts and of Indonesian Papua, is filled with instances of brute force on the part of settlers and the military protecting them, at times with gratuitous savagery, and sophisticated analysis seems beside the point. The only thing to be done is to mobilize opposition to these policies and to the international support or toleration that they receive.

There are, however, several reasons for not leaving the matter with mere description and for moving on to ethical analysis. One is the simple observation that the detailed facts often take a long time to appear on the stage of public attention. They typically relate to the dark areas of the globe as far as the spotlights of the mass media are concerned. State actions must not be left to go unchallenged until their claims regarding the facts of such cases are demonstrated to be false or misleading. In the meantime, certain state rationales can be challenged on the grounds of their moral unacceptability, quite apart from disputes about the precise facts of the case, such as the unconditional moral primacy of state sovereignty. Moreover, the relevance of facts does depend on the moral framework adopted in assessing them. If the world public or international lending and foreign aid agencies believe, for example, that state sovereignty is unconditional, then much information about what indigenist advocates perceive to be injustices will simply be treated as irrelevant to decision-making. When unjust policies are involved, state legitimation, which even authoritarian and wholly amoral regimes find it necessary to engage in domestically as well as in relation to the international community, has to be challenged, not only at the factual level, but also at the level of the moral categories and arguments employed. Not doing so increases the danger, constantly faced by social-justice advocates, that they fail to recognize the different moral reflexes of the public and of decision-makers and, by not addressing them, find themselves politically marginalized.

A second reason for the ethical analysis of policies is that political partisans, by treating opponent forces as evil, sometimes fail to recognize that there is a real ethical controversy, and thus a need for dialogue. In the process they may actually discover new facts and relevant considerations that they had neglected and refine their own position, just as this may happen to their opponents. Without being naive and proposing to substitute dialogue for struggle, it is important to recognize the role for both.[49]

The following evaluation of state rationales will not take the form of careful textual analysis or the examination of the details of arguments. Political rationales rarely stand up to critical scrutiny of this kind. Since they are not expected as communications with scholars, but as communications with the public, mostly through the crude medium of mass communication, to expect them to meet scholarly requirements of consistency and completeness of argument in public presentation is more than can be expected. Instead, I will identify certain general lines of justification and assess how well they stand up to critical evaluation. Moreover, my approach to ethical evaluation will take the form of an immanent critique. That is to say, I will begin with the developmentalist value framework and start the critical evaluation from within that framework, using concepts from mainstream economics. This will also mean that initially I put the most favorable interpretation on the policy rationale considered. That initial framework will then be extended step by step, bringing in concerns related to ecology, distributive justice, and freedom.

As a basis for classifying policy rationales, I will begin with the Indonesian government's declared goals for its transmigration policy, as articulated in the early 1980s. These are the promotion of national unity, national security, an equal distribution of the population, national development, the preservation of nature, assistance to the farming classes, and improvement in the condition of local people.[50] These various policy rationales can be distilled into the following simplified set: (1) national development, (2) environmental protection, (3) alleviating heartland poverty, (4) developing "backward" peoples, and (5) territorial integrity and national unity. Referring back to the list of Indonesia's transmigration goals, "the promotion of national unity" and "national security" are represented by category 5. "An equal distribution of the population" is covered in category 3. "National development" is represented by its own category (1) as is "the preservation of nature" under the somewhat broader notion of "environmental protection" (category 2). "Assistance to the farming classes" is also accommodated in category 3. "Improving the condition of local people" is covered by "developing 'backward' peoples" (4), with the latter phrasing used to bring out more clearly how this objective is widely understood, as shown below.[51]

National Development

"National development" is a very accommodating conceptual vessel; you can put many very different notions into it without a problem of fit. Literacy, productive capacity, entrepreneurship, military power, or democratic institutions can all be developed on a national scale. That, perhaps, is its appeal as a political slogan. Let us set aside here various possible forms of political and other kinds of social development, and concentrate on what is normally understood by "development" in the contemporary developmentalist context, which is *economic* development. Although economic development centrally involves production, it still can mean rather different things. It can focus on the production of particular things, such as those that meet basic needs or on machines and infrastructure, or for particular groups, such as the poor or those with purchasing power or relevant skills; or it can refer to the nation's productive capacity in a more open-ended sense. Setting aside distributive issues for the subsequent section, I will here focus on the latter. It is the normative idea that underlies the statement by the World Bank that transmigration in Indonesia "brings under-utilized land into intensive cultivation, raises agricultural production, and contributes more broadly to economic development in the Outer Islands."[52] Of course, this broad notion leaves much scope for dispute about the shape that national economic development should take and the purposes it should serve.

Let us consider two broad rationales for the development of the economy's productive capacity: (1) capacity to produce goods and services desired by those with purchasing power obtained from their productive contributions and from subsidies issued by the state on other grounds; (2) the ability to increase the previously mentioned capacity in the future, that is, economic growth. Under mainstream developmentalist ideology, these should be maximized (with certain qualifications), whatever the appropriate distribution of the benefits from production is. While there is clearly a tension between production for consumption now and for consumption in the future, the criterion requires that productive resources are not wasted or left idle (unless held for better future use), that people willing to work are not left unemployed and that unwanted things are not produced. In particular, in frontier areas, natural resources, such as minerals and timber, fertile land and labor should all be put to productive use. The productive use of labor will be discussed in the next sub-section. Here I will focus on development in the form of the more intensive use of natural resources.

In response to the general argument that national development is promoted by the more intensive use of their natural resources in the frontier lands, three points are to be made: (1) Subsistence production, even though it is not integrated with the national economy, is still part of the economy's productive capacity and can effectively meet the needs of those practicing it; (2) many practices involved in the colonization and resource exploitation of tribal lands are actually inefficient in terms of national economic development; (3) national economic development, as here articulated, is morally inappropriate if taken to be the sole aim of state policy.

(1) Subsistence producers are productive. They provide for their own consumption requirements. In assessing the utilization and development of production capacity, it is simply mistaken to consider only production for markets. This point is perhaps more of symbolic than substantive significance. All it amounts to is that, in cost-benefit analyses, for example, the subsistence production of tribals should count, just as the production of homemakers and of those who do repairs around their houses should count. Only in limited cases will it, by itself, change the conclusion that introducing the modern modes of production to tribal areas will increase productive capacity. If the latter is the criterion, certain justifications for modernization—that shifting cultivation is less productive than settled agriculture,[53] that it is unproductive to leave forests unharvested and to leave minerals unextracted, that participating in a larger system of division of labor and acquiring the skills for it is more productive than having more or less all productive activities occur within a single village, and that export crops often contribute more to overall production, once the imports obtained with the earnings are counted, than subsistence or local-market crops—maintain their plausibility.

(2) Nevertheless, there are instances where new forms of cultivation, new kinds of extraction or new activities are inappropriate—either in the context of the particular natural environment involved or in the manner in which the innovations are carried out—that the change from the old mode of production to the new reduces rather than extends productive capacity. Thus, it is argued by a scholar of agricultural practices that, "where land is

abundant and resources scarce, it is generally agreed [shifting cultivation] is an efficient and stable system that has sustained farm families for many generations."[54] Its efficiency is enhanced by the fact that it requires little cultivation and management once the crops have been sown. Sometimes the terrain is so unsuitable for annual cultivation that even settlers have resorted to shifting cultivation. However, settlers often do not have the necessary experience with the frontier environment and are not given the necessary advisory support so that they often use inappropriate methods or carry them out inappropriately, with resulting depletion of soil fertility and thus a reduction in the economy's productive capacity. Moreover, certain development activities, such as logging, mining, and damming, have such serious environmental side effects, in the form of deforestation, erosion, pollution, river flow disruption, flash flooding, malaria, and damage to fisheries, that it sometimes becomes questionable whether, in terms of productive capacity alone, benefits really exceed both internal and external costs, especially to subsistence producers.[55]

(3) The most important point to make, however, is that treating the full utilization and expansion of the nation's productive capacity as the only goal is morally quite inappropriate. Other relevant considerations concern the distribution of the benefits from the productive capacity, the preservation of environmental amenities, the protection of tribal peoples against destructive cultural change, and the freedom to make choices. These will be explored in the following sub-sections.

Environmental Protection

The goal of preserving nature would normally be expected to require restraint on the development of tribal lands. Tribal peoples have acquired a well deserved reputation for respecting nature and for engaging in practices that are sustainable. Thus, there has been considerable coincidence of tribal interests and the concerns of environmentalists and this has been reflected in important alliances between the two advocacy movements that represent them. To find developmentalist states including environmental protection as a goal may suggest nothing more than rhetorical concessions to Western states and international organizations that cannot ignore Western environmental movements, or it may represent a genuine concern for the environmental sustainability of economic development that is jeopardized by certain frontier development practices, such as clear-cut logging and agricultural methods that do not pay attention to the maintenance of soil fertility.[56] The focus of the current discussion, however, is on state rationales for the development and colonization of tribal frontier lands. Can environmental protection be used to justify certain aspects of the colonization and exploitation of tribal frontier lands?

Two kinds of rationale are encountered in discussions of these matters. One is that shifting cultivation is not only a relatively unproductive practice, but is or can be environmentally unsustainable; and this applies also to other practices, such as hunting and fishing. This argument does not really go beyond the preceding rationale of developmental efficiency since it merely involves the maintenance of the economy's productive capacity.[57] The other argument, which does go beyond economic development, is that the preservation of nature requires restraints on the tribal use of lands set aside for that purpose. What is the basis of each of these arguments?

(a) Generally, tribal peoples who are relatively unaffected by modern ways have been observed to practice subsistence production in a manner that is sustainable. Those that did not do so in the past must have discovered that they had to change to sustainable practices in order to survive or they must have ceased to function as on-going tribal societies, either because they disintegrated or because they turned to conquest and the subordination of other peoples and thus ended their tribal existence.

There are, however, at least three conditions that can today lead to economic activities by tribal peoples that are environmentally unsustainable. First, if colonial or developmentalist states displace tribal peoples from all or part of their traditional lands or restrict their use of these lands and force them to rely on a narrower resource base, then it is well possible that, given their knowledge base, they use the remaining or new resources available to them in an unsustainable manner. In the Chittagong Hill Tracts that may well have been happening after the construction of the Kaptai Dam and the subsequent displacement through Bengali settlement. Second, tribal peoples that have been significantly touched by the modern world are characteristically thrown off balance demographically. In response to modernizing contacts tribal populations tend to grow quite rapidly (at times following devastating declines due to disease upon initial contact) and this creates definite pressures on the environment.[58] Third, restrictions that colonial or developmentalist states have placed on tribal peoples' control of their natural environment (in order to appropriate natural resources for themselves or their privileged groups) make it impossible for tribal peoples to exercise full responsibility for their natural environment. As a result, they can lose the customary patterns that have served conservation in the past. In other words, the reputation for conscientious conservation may be more applicable to autonomous and unacculturated tribes than to those that have become entangled with the modern world.

In assessing such an argument, the first point to be made is that categorical generalizations are inappropriate here. Thus, just as indigenist advocates cannot validly claim that indigeneity in itself assures environmentally protective practices, so states cannot claim that indigenous practices in general are environmentally damaging. Moreover, even if such practices are unsustainable, the fact that this is due to deliberate actions by agents of modernization or due to their side-effects surely must mean that merely putting further restrictions on tribal access to natural resources is not acceptable, just as the status quo is not either. Obligations to tribal peoples as a matter of justice here become crucial. These will be discussed in the next section.

(b) The preservation of nature, in the strict sense in which it goes beyond merely sustainable use, can require that indigenous people are excluded from their traditional habitat. Thus, the OPM criticized the World Wildlife Fund for organizing a nature reserve that excluded not only new settlement and commercial resource extraction, but also controled Papuan access to resources and displaced 102 Papuan families.[59] The liberation movement concluded that the World Wildlife Fund was an enemy rather than an ally. Is an action such as that of the World Wildlife Fund justified? (I am here substituting the World Wildlife Fund for the Indonesian state as a more credible proponent of the goal of the preservation of nature, but one that must have collaborated with the Indonesian state to organize this nature reserve.) Clearly, the project was based on a goal other than the enhancement of Indonesia's or the world's productive capacity.[60] In the first instance it might be argued for in terms of the interests of future generations in the enjoyment of nature.[61] Such a value criterion, however, is still anthropocentric in that the value of nature derives from what human beings get out of it. Various schools of environmentalism insist that such an approach is "speciesist" and represents a lack of impartiality analogous to such a lack in racism or sexism. They argue for the transcendence of such "resourcism" and for the recognition of the inherent value and the rights of other species of life and of ecosystems.[62]

In response, it has to be pointed out that, however meritorious the preservation of nature for its own sake, there is an issue of justice with respect to the distribution of sacrifices to provide for it. Tribal societies in fact can make a case wholly analogous to that often made by Third World regimes. It is unjust for the highly industrialized countries, which have in the past disproportionately extracted nature's resources and disproportionately imposed on nature's capacity to absorb wastes in the process of making themselves disproportionately rich, to now ask the Third World to restrain its development activities for the sake of the common global interest. But if that is unjust, it is doubly so when Third World states shift this restraint to Fourth World societies. The solution, promoted by the United Nations, of biosphere reserves that incorporate traditional sustainable tribal practices rather than exclude them, seems an eminently reasonable solution. But this involves invoking criteria of justice and addressing the issue of tribal autonomy, both of which have yet to be brought into this discussion.

Alleviating Heartland Poverty

Beginning with distributive justice, I will first consider a rationale for the colonization of tribal frontier lands in terms of an obligation to share. The notion is that tribal peoples have much land and resources, while there are many people in the national heartland, that is, in the lowlands in Bangladesh and on the island of Java and certain nearby smaller islands in Indonesia, who are landless rural laborers or poor peasants whose poverty can be alleviated by settling them in these frontier areas. Moreover, the extraction of resources can also help finance development projects for the heartland poor. It is simply a matter of equalizing access to natural resources. Such sharing was, for example, appealed to by a diplomatic representative of the Bangladesh mission in Geneva, who, in a speech to the United Nations Working Group on Indigenous Populations concerning the Chittagong Hill Tracts, declared: "The people of Bangladesh have to share the available resources in an equitable manner within the severe constraints well known to the international community."[63] This idea also seems to be implicit in the statement by the World Bank that "transmigration can provide employment and land ownership to some of the poorest members of the labour force in Java."[64]

In response to this rationale, I will make the following case. Accepting, at least for the time being (to be reconsidered below), the principle of resource sharing within a territorial state, I will argue that it is not the heartland poor that are the relevant moral reference group here, but a much wider reference group. In the following sub-section, I will further argue that imposing a more intensive use of resources on tribal peoples can be destructive to their culture and the social-psychological supports they derive from it and that this further constrains claims that can be made on their resources. Finally, in a further subsection the principle of resource sharing will be considered more closely in the light of the issue of national sovereignty.

Formulating the issue in terms of distributive justice between the heartland poor and the frontier tribals is to accept an articulation by national elites, which, on closer inspection, turns out to serve their class interests. It is not only the comparison of the conditions of the poor in the national heartland and of tribals in resource-rich frontier areas that is relevant. There are also comparisons to be made between the better-off classes in the national society and their poor and also between those better-off classes and the tribals. In other words, we here have what I call a "three-sided justice issue," involving at least a three-way comparison. It is certainly self-serving and thus corrupt ethics for social groups to present the ethical issue in terms that exclude themselves even though crucial obligations fall on them when a more comprehensive assessment of distributive justice is made. In that case, what ought to happen is that the comparative assessment is made across all groups in the territorial state. If resource-rich tribal communities have obligations to share with the heartland poor so do certainly the heartland rich.[65]

Moreover, it is not only natural resources that are relevant to this assessment, but productive resources in general. It is true that, to some extent at least, the accumulation of productive resources is a function of past sacrifices and of judiciousness in responding to the consumption or production-input needs of others, which ought to be rewarded, if for no other reason than as incentive to productive effort. However, this is so entangled with acquisitions based on unequal inheritance without basis in merit that it simply cannot be excluded. Moreover, regarding distributive justice between the national heartland as a whole and tribal frontier lands, much productive capacity, whether in the form of buildings and machines or public infrastructure, has historically involved the process of converting natural resources into human-made resources; it thus reflects previously held but now used-up natural resources. Furthermore, the conservation of natural resources by tribal peoples is based on demographic self-restraint in contrast to the demographic profiligacy of the national heartland; this too is entitled to reward.[66] Counting productive resources as a whole means clearly that in only rare instances will tribal peoples be found to be inordinately rich in resources, given the historical process of displacement of tribal peoples from valuable lands by expansive societies that occurred before the current era of developmentalism. In fact, it will generally be tribal peoples that will have a claim to redistribution in their favor.

Even if, in a few cases, tribal peoples happen to be endowed with natural resources that world demand has made very valuable, there are further reasons for constraints on redistributive claims by the national heartland. These are, first, the culturally destructive effects of the alienation of those resources and of being forced into the modern economy without resources for negotiating the terms of such entry, and, second, rights to self-determination that distinct peoples or nations can claim. These will be considered in the remainder of this section.

Developing "Backward" Peoples

If distributive justice claims can be made on behalf of the heartland poor, they can also be made on behalf of the tribal peoples—and they have, even by developmentalist states. However, the way that such claims are typically incorporated in policy goals is in the form of bringing the benefits of development to "backward" tribes.[67] Tribalism and backwardness are equated by definition in Indonesia's 1945 constitution where tribal peoples are those groups that inhabit remote communities "whose social life, economic performance and level of civilization are below acceptable standards."[68] More recently, an Indonesian representative, speaking before the United Nations Working Group on Indigenous Populations, declared:

> The objective of Indonesia's transmigration programme, which is a national endeavour already in existence for many years, is to expand development efforts and to evenly spread its benefits to the regions outside of the already overpopulated areas in order to achieve nationally balanced economic progress. The purpose is to utilize the surplus agricultural manpower available to develop land resources in the outer islands. It is aimed at improving the standard of living of the community in general, by increasing regional development and by assisting the people on the outer islands that demonstrate a relative lag in development.[69]

Bringing development and its benefits to tribal peoples raises two issues. They are: (1) what happens to the traditional entitlements of tribal peoples to their natural resources in the development process, and (2) the impact of development on tribal culture.

(1) What would development look like if it was in fact guided by the aim of benefiting tribal peoples? How would it avoid the following charge made by one advocate for the tribal peoples of the Chittagong Hill Tracts?:

> To the Bengalis, development has meant the acquisition of the natural and human resources of the Chittagong Hill Tracts. To the Tribal People, however, development has meant the deprivation of communally owned resources, exploitation of their lands, eviction from their homeland and, finally, genocide.[70]

The fact that the state in both Bangladesh and Indonesia declared eminent domain over lands without private-property titles suggests that the approach to the development of tribal peoples, to the extent that it is genuine at all, has been intended to take a quite coercive form. Tribal peoples are left without bargaining power of any kind in the development process. It means that development becomes what the state and other agents of modernization decide it should be. That it led to militarization in both the cases under analysis is simply a reflection of the coerciveness of this process.

If, on the other hand, the aim was simply to enhance the productive capacity of tribal peoples for their own benefit, so that they could then use this enhanced productive capacity to increase their range of choice, the latter being one way in which the rationale of development in general is sometimes articulated, then that implies giving tribal peoples first title to their lands and the resources on or under it and then helping them find ways to make more productive use of them themselves or to use them to strike beneficial bargains with agents of development such as corporations. Development initiated through expropriation seems a perverse and, in fact, downright dishonest way of proceeding to bring the benefits of development to tribal peoples.

Two arguments in defence of such an approach to development might still be advanced. One is that this kind of development actually involves an exchange. The state and the agents of the national economy offer technology, entrepreneurship, employment opportunities, and markets for tribal producers and consumers in exchange for tribal relinquishment of natural resources. But it is not an exchange that tribal peoples freely enter into; it is foisted on them.

The other argument is that, if indigenous peoples retain control over their lands and resources, they will not want to develop, simply because of their "primitive" outlook, and will thus not do what is in their own best interest. Such paternalism, however, violates certain choice requirements of economic efficiency. The latter requires not only the avoidance of waste in the use of productive resources, but also that it is up to individuals to determine what they deem to be the benefits of potentially economic goods and factors. If tribal people consider it more important to maintain certain sites as sacred rather than use them for cultivation, then that means, in terms of the efficiency logic, that the value to them of the site in its sacred "use" is greater than the value of any other use.[71] In fact, such developmentalist paternalism has proven to be harmful to tribal peoples in the past. One reason for this is a lack of recognition of the importance of their culture to the well-being of tribal peoples, which will be briefly expanded on next. The general point here though is that, contrary to the political ideology that has historically been associated with the economic ideology of development, the ideal of freedom here is totally negated.

(2) Being suddenly thrust into the system of modern economic relations is typically destructive to tribal cultures and the social-psychological supports they offer to their members. The concern here is with the well-being of tribal people, not with the preservation of social-structural systems for their own sake. Quite apart from the limited knowledge that tribal peoples have about the modern economy and its opportunities, pressures and hazards, human beings are not so adjustable that they can readily accommodate themselves to the shock-like transformation that is involved.

One common consequence that has been observed is a serious deterioration in health conditions resulting from the loss of land providing traditional foods, the disruption of customary trade systems, relocation to new lands with unfamiliar sources of food, dependence on unreliable or exploitative outside employment and debt-bondage, the adoption of new clothing practices without the hygiene required for it, or a change to cheap or prestigious processed foods.[72] Quite apart from the change-over in modes of production that are involved, modernization, especially if it involves giving up land, requires the abandonment of certain traditional attitudes towards land, such as its endowment with value on the grounds of religion, ancestral significance, or the location of particular tribal customs. Even without displacement, the initial impact of Western civilization has been observed to have an enervating effect on indigenous communities. If the communities persist, they often do so in a relatively depressed state.[73] The social-psychological process that is involved appears to consist of a loss of self-esteem due to the denigration of tribal culture in developmentalist propaganda, the loss of a variable, highly self-determined and leisure-rich pattern of living in exchange for one characterized by the discipline of settled agriculture or employment, and the development of aspirations for affluence that prove to be unattainable for tribal people in the real world of the modern economy, even for those willing to adapt.[74]

If the tribal communities do not persist and their survivors drift off to the towns and cities of the national society, they typically end up on the periphery of those communities and come to form a new underclass or join an existing one.[75] In short, a transformation that is supposed to benefit tribal peoples turns out to have the very opposite effect.

These points regarding property titles and cultural protection also have implications for the preceding discussion regarding the distributive-justice obligations of resource-rich tribal peoples with respect to other groups in the territorial state. First of all, even if there are disproportionately resource-rich tribal peoples who, considering the matter only in terms of productive endowments, have redistributive obligations to others in that system, there is no reason that those obligations should be met by state appropriation of the natural resources of tribal societies. Resource taxation, which leaves it to the resource-rich tribe to determine how they raise these taxes, is sufficient. Secondly, the case for such obligations must be qualified by the culturally destructive consequences that requiring resource access to outsiders would have. It may well be that redistributive obligations, if they are not to have the effect of making their holders worse off than everyone else in the system by destroying their culture, can only be phased in over a long enough period that manageable adjustment is possible. And that may well involve two, three or four generations. In other words, tribal peoples have to be allowed to continue, for a considerable period, to use more extensive resources in their traditional, less intensive manner if equality, rather than the exploitation of tribal peoples, is the aim.

Territorial Integrity and National Unity

Two further purposes of the colonization of frontier lands that are ostensibly appealed to are those of maintaining territorial integrity and promoting national unity. In fact, the Bangladesh and Indonesian regimes have interpreted these aims in terms requiring state sovereignty in very centralist terms and, being suspicious of the loyalty of the populace in some of their frontier regions, have pursued strategies of "nation-building" that involve the denial of the distinctive national identities of their frontier populations. In Indonesia, despite its national slogan of "unity in diversity," the minister of transmigration in 1985, Martono, declared that:

> by way of Transmigration, we will try to realize what has been pledged, to integrate all ethnic groups into one nation, the Indonesian nation...The different ethnic groups will in the long run disappear because of integration...and...there will be one kind of man.[76]

In Bangladesh, at the time of its founding, a proposal for making the Chittagong Hill Tracts an autonomous tribal region during the debates of the Constitutional Assembly was considered a conspiracy against the sovereignty of Bangladesh and tantamount to secession;[77] as previously mentioned, the first President of Bangladesh advised a tribal delegation to replace their traditional "ethnic" identities with a commitment to Bengali nationalism. In fact, all of Bangladesh's presidents have "always defended military suppression of the peripheral ethnic communities on the plea of 'national integration'."[78] Part of this commitment to the assimilation of minority nationalities to the dominant nation presumably stems from strategic and security considerations, not really with respect to the safety of the national heartlands, but with regard to the territorial integrity of the periphery.

Territorial integrity is maintained in these cases by pursuing national unity through the denial of political processes for consent by the frontier peoples, simply because they are suspected of disloyalty as a result of their commitment to distinct national identities. Natural solutions would have been a federal system in the case of Indonesia, given its great national diversity, and of an autonomous region in the case of Bangladesh, if there is no need for a federal system on other grounds.[79] Unfortunately, in both cases the predominance of one national group and its reluctance to relinquish power and direct control over the natural resources of frontier lands have prevented such a solution.[80] Instead, territorial integrity is pursued by demographic subversion—that is, turning the indigenous population into a minority in its own territory, and by military coercion. Since "national unity" here is essentially a means to the end of territorial integrity, I will focus the ethical evaluation on the latter goal exclusively. Under the conditions prevailing in Indonesia and Bangladesh and given the means employed there, is there a rationale for such maintenance of territorial integrity, so that it should receive the consent and support of the international community?

The first issue is the manner in which territorial integrity is maintained. Accepting for the moment the moral primacy of territorial integrity, it has to be explored what other moral requirements can be met within those terms of reference. In particular, there is the fundamental principle of self-determination. While this is a principle that has been well legitimated in the international community and in particular through the United Nations, it has, at the official level, not been articulated either with respect to (1) who is entitled to claim it or (2) what it requires.[81]

(1) The United Nations General Assembly Resolution 1514 states that "all peoples have the right to self-determination."[82] Regarding the interpretation of "peoples," one position is that "a people is a distinct ethnic group, the identifying marks of which are a common language, shared traditions, and a common culture."[83] Even if one interprets "peoples" more restrictively as "nations," where the latter are defined not only by the features of ethnic groups but also by having had experience with political self-determination at least in the past,[84] then the peoples of both the Chittagong Hill Tracts and of Indonesian Papua clearly qualify. Historical experience need not have been with centralized self-determination, but may have taken the form of small-scale (even village) polities that had independent decision-making power of a substantial kind.

(2) With respect to content, the principle of self-determination may range from as strong a form as what Allen Buchanan has called the "normative nationalist principle [which]...states that every 'people' is entitled to its own state"[85] to the much more limited requirement that there be a political jurisdiction and process for a people to determine certain crucial matters concerning its governance. Considering the latter in order to remain for now within the framework of territorial integrity, what is crucial for the limited structures of self-determination is that the indigenous peoples can adequately protect their well-being, in the light of their distinctive vulnerabilities. That means that they must have title to their lands and their resources, including forests and minerals, and they must have it in a form that cannot be readily revoked by the central government; they must have collective control over the access of outsiders to the particular region and over the sale and leasing of land and resources to outsiders; and they must have control over the legal system and the enforcement of laws as they relate to the matters mentioned.[86] This does not mean that a central government would have no jurisdiction over the region whatsoever. Certain powers, such as foreign relations and defence,

country-wide and inter-state economic relations, and certain kinds of taxation would naturally remain in the hands of the central government. However, resource exploitation and colonization is something that would have to be negotiated with and consented to by the one or more governments of the tribal region. Rather than being a threat to territorial integrity, a federal structure is precisely what can provide for "unity in diversity" (as it has in many countries), by avoiding the resentment and resistance generated by the attempt to enforce assimilation and to impose economic management from outside, quite apart from the expropriation of land and resources.

Secondly, can tribal self-determination rights trump the right to maintain territorial integrity by existing states? A review of Buchanan's careful discussion of the right to secede,[87] the limits to it, and justifications for the forcible prevention of secession suggest that, in the cases at hand, there is no sustainable argument for denying self-determination in the form that can include forming an independent state or joining an adjacent one (as some groups in West Papua favor with respect to Papua New Guinea).

On the other hand, there are several reasons in favor of a moral entitlement to secession in these cases. The tribal peoples have not been consulted under conditions of free choice about the accession of their respective regions to the larger states. In the case of the Chittagong Hill Tracts they were incorporated in 1947 against their wishes, as expressed by their leadership, by an agreement negotiated between a departing colonial power and two successor states that did not represent their interests. In the case of Dutch West New Guinea, the consultation in 1969 occurred under conditions of coercion and without adequate representation. This violates not only the principle of consent, but represents a justification for secession on the grounds of "rectificatory justice."[88] Moreover, this means that, even if it were to be found that these tribal regions as a whole were inordinately resource-rich relative to the state heartlands, the requirement for a distributively just secession settlement with respect to resources would not hold in this case, because their incorporation in these territorial states was unjust in the first place. Distributive justice would have to be assessed as a matter between (potentially) sovereign states. It is only to the extent that international distributive justice is being implemented in general that distributive-justice obligations fall on regions whose legitimate right to secession has been denied to them.

This historical background also is decisive in dealing with the complication of substantial state-heartland populations now residing in the tribal regions; since their influx occurred after the time when proper consent ought to have been obtained and since it was part of a deliberate policy of demographic subversion, these settlers are there illegitimately and it must be left up to the tribal people as to whether they must return or can stay. While this may involve hardship for these settlers, the obligation for facilitating their return and for any compensation owed to them falls on the expansionist states and the foreign powers and international institutions (for example, the United Nations in the case of West Papua) responsible for the initial injustice. If recent influxes of settlers were allowed to undercut the right to secession, there would be a clear incentive to territorial states that have border regions populated by nationalities with questionable loyalty to swamp them with loyal heartland groups. As far as state investment in the frontier regions are concerned, these were not chosen by the indigenous people and typically served purposes hostile to their interests. If compensation is owed, it is by the developmentalist states to the tribal peoples for the resources extracted and the harm inflicted.

Apart from the injustice of the historical accessions, the subsequent treatment that the indigenous peoples received at the hands of the controlling states justifies secession. The expropriation of their resources and lands constitutes a justification for secession in terms of "discriminatory redistribution."[89] The states' attempt to destroy their distinctive ethnicity and culture constitutes another. Finally, secession is justified by the rampant violation of human rights and a policy that at least has verged on genocide as these states pursued their coercive strategies.

The OPM is thus on solid moral grounds in demanding from the United Nations "external self-determination."[90] The former colonial regimes of the British and the Dutch have simply been replaced by what has come to be called the "internal colonialism" of Bangladesh and Indonesia. Their elites, "once in charge of governments, have forgotten their own struggles and have used the tools of repression borrowed from their former adversaries."[91]

Summary

Five general rationales for the state policy of colonizing tribal lands and "developing" their resources have been considered. They have been (1) national development, (2) environmental protection, (3) alleviating heartland poverty, (4) developing "backward" peoples, and (5) territorial integrity and national unity. Neither on their own nor together have they been found to be convincing.

(1) Initially, with respect to the appeal to national development, only the cautionary points were raised in that subsistence production was to be recognized as part of the economy's productive capacity, that many development practices in tribal frontier regions are in fact economically inefficient and that national economic development is not the only relevant moral criterion. (2) To the extent that indigenous people end up engaging

in activities that damage the environment, this is due to the impact of modernization on them, including the resulting restrictions on their resource base, and whether it is they who should bear the burden of further restrictions to protect the environment becomes an issue of distributive justice. (3) Concerning the appeal to finding land and resources to alleviate the poverty of the heartland poor and to the distributive-justice obligations of tribal frontier peoples vis-à-vis these poor people, it has been argued that the reference group for assessing whether the tribal people are inordinately resource-rich is not the heartland poor, but in the first instance the heartland society as a whole, including the rich, counting not only natural resources but productive resources in general. (4) The rationale in terms of developing "backward" peoples runs into the objection that the pattern of development is one of coercion and expropriation as well as cultural destruction. With respect to the previous rationale this means that the distributive-justice obligations of inordinately resource-rich tribal groups must be such as to maximize autonomy and must be phased in over a period spanning generations. (5) Treating national unity as instrumental to territorial integrity in the cases under review, it has been argued here that the indigenous peoples involved have not only rights of significant self-determination within a federal or analogous structure, but, by virtue of the way their regions were incorporated into Indonesia and Bangladesh and of the exploitative, culturally destructive, and probably genocidal treatment to which they have been subjected, have the clear moral right to secession.

But is all this a surprise? Should not this have been expected all along? Was this ethical evaluation in fact worth doing? While we might have expected this conclusion, it was still worth going through the steps of the argument. Even if there were no need for social-justice advocates to question the explicit or implicit elements of their own arguments, merely exposing the lies and half-truths of Machiavellian states is not sufficient. It is necessary to engage the moral component of state legitimation, even if it is nothing but rationalization. One reason is that the precise facts are often slow in getting onto the stage of world attention. Another is that the state rationales have shown themselves to be acceptable enough to various publics and, in particular, to donor and lending agencies, despite active fact-finding by advocacy groups, so that the unjust policies have been sustained over a long period of time; it is possible that it is the legitimated moral framework, such as the moral primacy of state sovereignty, that allows such agencies to dismiss revealed facts in favor of considerations stemming from power relations. Finally, by not engaging moral notions different from their own, advocates may marginalize themselves politically and remain unheard. In other words, in addition to the important task of revealing what is actually going on, there is the task of debunking faulty justifications that serve simply as rationalizations. Indigenist advocates have, in fact, been doing some of this. The arguments provided here are intended as a contribution to this effort.

Concerning the reality behind the justifications that have been analyzed, I can do no better than quote Bernard Nietschmann's forceful words:

> Third World colonialism has replaced European colonialism as the principal global force that tries to subjugate indigenous peoples and their ancient nations...
>
> Invasion and occupation of indigenous nations once done by white expansionist powers are now done by foreign brown expansionist powers. What is called "economic development" is the annexation at gun point of other peoples' economies. What is called "nation-building" is actually state expansion by *nation-destroying*. Territorial consolidation, national integration, the imperatives of population growth and economic development are phrases Third World states use to cover up the killing of indigenous nations and peoples.[92]

Notes

1. I gratefully acknowledge help from Petrina Arneson, Mahnoor Yar Khan, Susanna Devalle, Jamil Rashid, Howard Adelman, Rhoda Howard, Alec Gordon, and Victoria Heftler. Part of the impetus to this research also comes from the shock of hearing, through the paper that Aditya Kumar Dewan delivered in 1988, about the guerrilla war in the Chittagong Hill Tracts, where I had lived as a teenager for several years in the 1950s, in the paper mill community briefly referred to below.

2. Although in some parts of the world the term "tribal" is deemed derogatory, I am using it nevertheless for the following reasons. (i) Alternative terms, such as "indigenous," "autochthonous," "Fourth World," or "First Nations," one or the other of which I will use occasionally, in themselves fail to capture some crucial features of the way of life of the peoples of concern here. While it emphasizes aboriginality, what really counts in the analysis being presented is whether the particular groups have remained relatively unintegrated with the mainstream economy of the country, are relying on less intensive modes of production, that is, use nature in a manner less mediated by technology, typically avoid the private ownership of land, and have a distinct culture. (ii) The terminological strategy of avoiding stigmatization seems to me futile; unless there is deeper change, over time the new term simply becomes stigmatized in turn. It strikes me as at least an equally promising strategy to assert the traditional terms with respect and, in the case of self-designation, with pride. Robert Anderson and Walter Huber, in *The Hour of the Fox* (New Delhi: Vistaar Publications, 1988), p. xii have concluded, with respect to the Indian context, that the term "tribal" "carries less and less pejorative value, but only insofar as the additional qualifiers 'primitive' or

'backward' have been eroded by the deliberate impositions of assimilation." (iii) At least in the Asian context, "tribal" is quite accepted in academic circles and is used by writers who are very respectful of tribal culture.

3. Bernard Nietschmann, "Economic development by invasion of indigenous nations: Cases from Indonesia and Bangladesh," in *Cultural Survival Quarterly*, Vol. 10, No. 2, 1986, p. 3.

4. For a similar use of this term, see Peter S. Wenz, *Environmental Justice* (Albany, NY: State University of New York Press, 1988).

5. Nietschmann, "Economic development by invasion," pp. 3-5; Syed Aziz-al Ahsan and Bhumitra Chakma, "Problems of national integration in Bangladesh: The Chittogong Hill Tracts, in *Asian Survey*, Vol. 29, No. 10, 1989, pp. 960-61; Aditya Kumar Dewan, "The indigenous people of the Chittagong Hill Tracts: A case study of violations of human rights in Bangladesh," (Paper presented at the annual meeting of the Canadian Sociology and Anthropology Association, University of Windsor, Ontario, 1988); Julian Burger, *Report from the Frontier: The State of the World's Indigenous Peoples* (London: Zed Press, 1987), pp. 129-30; Derek Davies, "Bangladesh 1: Four rays of hope: Self-rule offer to end 20 years of bloody conflict," in *Far Eastern Economic Review*, Vol. 143, No. 12, 1989, p. 20; Wolfgang Mey, ed., *Genocide in the Chittagong Hill Tracts, Bangladesh* (Copenhagen: International Work Group for Indigenous Affairs, 1984), p. 6.

6. Dewan, "The indigenous people of the Chittagong Hill Tracts," pp. 9-10, 13-16; Ahsan and Chakma, "Problems of national integration in Bangladesh," pp. 963-65; Mey, *Genocide in the Chittagong Hill Tracts*, pp. 24, 27, 104-06; Burger, *Report from the Frontier*, pp. 132-33; Nietschmann, "Economic development by invasion," p. 6; Akram H. Chowdhury, "Self-determination, the Chittagong and Bangladesh," in David P. Forsythe, ed. *Human Rights and Development: International Views* (New York: St. Martin's Press), p. 297; John H. Bodley, *Victims of Progress* (Mountain View, CA: Mayfield, 1990), p.10; Davies, "Bangladesh 1," p. 20.

7. Chowdhury, "Self-determination," p. 297.

8. In 1981, both the Australian Development Assistance Bureau and the Swedish International Development Authority withdrew from development projects because these projects were found to harm the tribespeople.

9. Ahsan and Chakma, "Problems of national integration in Bangladesh," pp. 965-68; Dewan, "The indigenous people of the Chittagong Hill Tracts," pp. 1, 11-13; Chowdhury, "Self-determination," pp. 297-99; Mey, *Genocide in the Chittagong Hill Tracts*, pp. 28, 65, 104-08; Bodley, *Victims of Progress*, pp. 10-11; Burger, *Report from the Frontier*, pp. 132-33; Nietschmann, "Economic development by invasion," pp. 6-7; "Bangladesh: insurgency in the hills," in *Economic and Political Weekly*, Vol. 24, No. 37, 1991, p. 2125; Davies, "Bangladesh 1," p. 20; Bimal Bhikku "Bangladesh: Chittagong Hill Tracts," in *IWGIA Newsletter*, No. 60/61, 1990, p. 23) has mentioned "more than half a million non-Tribal Muslims" for the end of the 1980s, Chowdhury "Self-determination," p. 299 has reported that "at the time of writing some observers estimated the two groups to be about equal in numbers," and Dewan ("The indigenous people of the Chittagong Hill Tracts," p. 1) estimated for 1987 equality between indigenous people and Bengali at 600,000 for each group.

10. Davies, "Bangladesh 1," p. 21; Ahsan and Chakma, "Problems of national integration," p. 968. Nietschmann ("Economic development by invasion," p. 7) referred to 15,000 guerrilla fighters.

11. Davies, "Bangladesh 1, p. 21; Chowdhury, "Self-determination," p. 298. There have also been factional killings within the Shanti Bahmi, including of its founder in 1986.

12. Mey, *Genocide in the Chittagong Hill Tracts*, p. 132.

13. Bhikku, "Bangladesh," p. 28; and see Dewan, The indigenous people of the Chittagong Hill Tracts," p. 12. The much lower figures of 30,000 and 40,000 have been given in Burger (*Report from the Frontier*, p. 133) and Nietschmann ("Economic development by invasion," p. 6), respectively. Bhikku, however, has provided a breakdown in terms of military units.

14. Dewan, "The indigenous people of the Chittagong Hill Tracts," pp. 16-18; Bhikku, "Bangladesh," pp. 25-26.

15. Burger, *Report from the Frontier*, p. 133.

16. In 1988, Bangladesh adopted Islam as its state religion.

17. Bodley, *Victims of Progress*, p. 11; Bhikku, "Bangladesh," pp. 27-28.

18. Dewan ("The indigenous people of the Chittagong Hill Tracts," p. 13), who is personally familiar with that region, committed himself, in his carefully referenced paper, to no more than that "thousands of people were killed" and Nietschmann ("Economic development by invasion," p. 7), who referred to the military-assisted settler invasion of the Chittagong Hill Tracts as "the world's most clear-cut example of genocide in practice today" did not offer a figure of any kind.

19. For example, Dewan, "The indigenous people of the Chittagong Hill Tracts," p. 13.

20. "Chakma refugees, in *Indian Express* (Kochi: 27 October, 1991, p. 7).

21. The refugee figure for Tripura that was most widely used in the late 1980s was about 50,000 (see Davies, "Bangladesh 1," p. 21; Dewan, "The indigenous people of the Chittagong Hill Tract," p. 13; and Mey, *Genocide in the Chittagong Hill Tracts*, pp. 33-35). The magnitude of these figures has been challenged by the Bangladesh government, according to which it was less than 30,000 at its peak, with about 8,000 subsequently returning (Davies, "Bangladesh 1," p. 21.) On the other hand, Bhikku ("Bangladesh," p. 25) has referred to "90,000 Tribal refugees...in 6 refugee camps in the Tripura State in India." The refugee situation of Chakmas in the state of Mizoram is unclear because they are not welcome in that state, where there is tension between the Mizos and the Indian Chakmas (personal communication from Mahnoor Yar Khan), and some earlier refugees from the Chittagong Hill Tracts appear to have been forcibly returned, but "according to one source [dated October 1986], there are 60,000 Chakmas who have settled in Mizoram, and most of them are refugees" (Mey, *Genocide in the Chittagong Hill Tracts*, pp. 33, 47-48).

22. Mey, *Genocide in the Chittagong Hill Tracts*, pp. 160-62.

23. Burger, *Report from the Frontier*, p. 134; Davies, "Bangladesh 1," p. 21; Chowdhury, "Self-determination," p. 298; Ahsan and Chakma, "Problems of national integration in Bangladesh," p. 970.

24. Nietschman, "Economic development by invasion," p. 7.

25. Davies, "Bangladesh 1," p. 21.

26. Davies, "Bangladesh 1," pp. 20, 21.

27. Bhikku, "Bangladesh," p. 23.

28. *Bangladesh: Reprisal Killings of Tribal People in the Chittagong Hill Tracts in May 1989* (London: Amnesty International, 1989, p. 2).

29. "Bangladesh: Insurgency in the hills," p. 2125.

30. Mariel Otten, *Transmigrasi: Myths and Realities: Indonesian Resettlement Policy, 1965-1985* (Copenhagen: International Work Group for International Affairs, 1986), p. 1.

31. Graeme J. Hugo, T.H. Hull, V.J. Hull, and G.W. Jones, *The Demographic Dimension in Indonesian Development* (Singapore: Oxford University Press, 1987), pp. 179-83. Nietschmann ("Economic development by invasion," p. 7) has referred to five million persons for the 1985-89 plan period, which assumes a rather large average size of family.

32. Malcolm Gault-Williams, "Strangers in their own land," in *Cultural Survival Quarterly*, Vol. 14, No. 4, 1990, p. 46; Sue Roff, "Multilateral development assistance destabilizes West Irian," in *Cultural Survival Quarterly*, Vol. 10, No. 1, 1986, p. 40; Nietschmann, "Economic development by invasion," p. 7.

33. Burger, *Report from the Frontier*, pp. 142-43, 147; Nietschmann, "Economic development by invasion," p. 7. This is in violation of the 1980 United Nations *Plan for Action for the Full Implementation of the Declaration of the Granting of Independence to Colonial Countries and Peoples*, Article 8, which deals with changes to the demographic composition of territories under colonial domination, over which particularly the Scandinavian members of the World Bank have expressed concern, as well as of the World Bank's own guidelines with respect to the consent of tribal peoples (Roff, "Multilateral development assistance," pp. 40, 41).

34. Nietschmann, "Economic development by invasion," p. 7; Gault-Williams, "Strangers in their own land," p. 46. Perhaps this astronomical figure, even if it does appear in one or more long-range plans, should not be taken too seriously since transmigration aims have been notoriously overambitious and unfulfilled in the past. Hugo, *et al.* (*The Demographic Dimension*, pp. 309-10, 330, 352-53) have used even for the high transmigration projections figures that are below the plans of the Department of Transmigration. Unfortunately they did not provide the anticipated transmigration figures themselves, only the figures for regional population growth.

35. Burger, *Report from the Frontier*, p. 145.

36. According to Nietschmann ("Economic development by invasion," p. 9), "the goal of the Pemka and Viktoria branches of the OPM is to resist the Indonesian invasion as long as it is necessary until world opinion, the cut off of international funding, or future allies pressure a withdrawal."

37. "Although many countries were greatly dissatisfied with [Indonesia's] handling of the purported plebiscite…, they also demonstrated a certain willingness to accept the situation because there was no clear evidence of an independence movement" (Roff, "Multilateral development assistance," pp. 40-41.) Thirty member countries abstained from the resolution to "take note" of Indonesia's

"Act of Free Choice." According to Roff (p. 41), decolonization law has not been brought into play because the "threshold" principle that requires evidence of sufficient numbers supporting independence has not been deemed to have been met.

38. "*Irian*" acronym for *Ikut Republik Indonesia Anti-Netherlands* ("follow Indonesia against Holland"), an Indonesian slogan from the campaign to acquire West New Guinea. From *Iryan*: "steamy land rising from the sea" (Biak language); *Jaya*: "victorious" (Indonesian)" (Gault-Williams, "Strangers in their own land," p. 48n1.)

39. Gault-Williams, "Strangers in their own land," pp. 44-45.

40. Burger, *Report from the Frontier*, p. 146; Carmel Budiarjo, *West Papua: The Obliteration of a People* (London: TAPOL, 1984, p. 34).

41. Gault-Williams, "Strangers in their own land," p. 45; Burger, *Report from the Frontier*, p. 146; Otten, *Transmigrasi*, p. 171; Budiarjo, *West Papua*, p. 38; J.M. Hardjojo, *Transmigration in Indonesia* (Kuala Lumpur: Oxford University Press, 1977, pp. 39-40).

42. Gault-Williams, "Strangers in their own land," pp. 45-46; Burger, *Report from the Frontier*, pp. 20, 146.

43. Organisasi Papua Merdeka, "Indonesia: Statement before the U.N. Working Group on Indigenous Peoples," in *IWGIA Newsletter*, No. 62, 1990, p, 90. The more open-ended estimate given by Gault-Williams ("Strangers in their own land," p. 46) is between 400,000 and one million.

44. Otten, *Transmigrasi*, p. 165.

45. Gault-Williams, "Strangers in their own land," p. 43; Bodley, *Victims of Progress*, p. 56.

46. The refugee figure for Papua New Guinea is from Organisasi Papua Merdeka, "Indonesia statement," p. 89; and Burger, *Report from the Frontier*, pp. 146-47. Burger reported more than 6,000 by the end of 1984 and more than twice that by the end of 1986. Bernard Nietschmann and Thomas J. Eley ("Indonesia's silent genocide against Papuan independence," in *Cultural Survival Quarterly*, Vol. 11, No. 1, 1987, p. 77) reported 15,000 by late 1985. As late as November 1988 they still did not appear in official refugee statistics; see C. Kismaric, *Forced Out: The Agony of the Refugee in Our Time* (New York: Human Rights Watch/W. Morrow, 1989), 119. In spite of widespread sympathy for the asylum seekers among the population of Papua New Guinea, the government there has been reluctant to extend such recognition, because it has been worried about its relations with Indonesia. In fact, it has committed itself to returning the "illegal border crossers" and has actually begun to do so, with cited cases of imprisonment and killings in Indonesia Papua following forced repatriation. In addition, it has taken measures to prevent crossings in the first place. It has thus violated international norms in the treatment of refugees.

47. Burger, *Report from the Frontier*, p. 143.

48. Steven Lukes, *Marxism and Morality* (Oxford: Oxford University Press, 1985, p. 1 and ch. 3.

49. Sometimes dialogue may be the only effective form of struggle available. When regimes are very powerful and do not have to worry about losing domestic and international legitimacy for the time being, there may be only one remaining

route for advancing morally desirable policy changes and that is to give decision-makers, at least on a trial basis, the respect of treating them as moral agents and at the same time to impress on them flaws in their moral position. After all, regimes are staffed by human beings who need to justify their actions also to themselves and that process of self-justification, if not impregnably protected by cynicism or an image of their (morally?) proper role as totally amoral, may be susceptible to moral argument.

50. Nietschmann, "Economic development by invasion," p. 7.

51. In another paper (Peter Penz, "Development refugees and distributive justice: Indigenous peoples, land, and the developmentalist state," in *Public Affairs Quarterly*, forthcoming) I deal with a number of these issues in a philosophically more elaborated manner, critiquing the indigenist rationale for morally fundamental (as opposed to legal) autonomy and arguing instead for global egalitarianism, but appropriately constrained by some of the considerations mentioned in the discussion below, including that of tribal autonomy.

52. *Report on Indonesia* (Washington, DC: World Bank, 1985), cited in Otten, *Transmigrasi*, p. 5.

53. See, for example, F. Kasryno, N. Pribadi, A. Suryana, and J. Musanif, "Environmental management in Indonesian agricultural development," in D. Erocal, ed., *Environmental Management in Developing Countries* (Paris: Organization for Economic Cooperation and Development, 1991), pp. 164-65; and the discussions of the attitudes of colonial and post-colonial authorities and contemporary consultants in Mey, *Genocide in the Chittagong Hill Tracts*, pp. 74, 104, Bodley, *Victims of Progress*, p. 10; and Walter Fernandes, G. Menon, and P. Viegas, *Forests, Environment and Tribal Economy: Deforestation, Impoverishment and Marginalisation in Orissa* (New Delhi: Indian Social Institute, 1988), p. 67.

54. Miguel A. Altieri, *Agroecology: The Scientific Basis of Alternative Agriculture* (Boulder, CO: Westview Press, 1987), p. 79.

55. For such effects in Indonesian Papua and in the Chittagong Hill Tracts, see Otten, *Transmigrasi*, p. 11, 22, 89, 94; Burdiarjo, *West Papua*, p. 38; and Mey, *Genocide in the Chittagong Hill Tracts*, p. 84.

56. Thus, Nietschmann ("Economic development by invasion") has mentioned, with respect to Indonesian Papua specifically, "deforestation and soil damage at the rate of some 200,000 hectares per year (to total 3,600,000 deforested hectares by 1989)" (p. 7), and, with respect to Indonesia as a whole, "environmental damage on a geographic scale verging on [that of] the 1941-45 War of the Pacific" (p. 3).

57. The claim that tribal practices are environmentally destructive can also mean that they are destructive to the environment as it is required by some other resource use, such as commercial logging. However, that rationale collapses completely into the preceding national-development argument. It becomes simply a question of whether tribal subsistence agriculture or commercial logging is more productive. More generally, environmental protection can be directly related to productive capacity. As "natural capital," nature is part of this productive capacity, or rather its resources are. To the extent that development projects impair these natural resources, such as inundate fertile land (e.g., the Kaptai Dam), erode productive soil (e.g., hillside logging in both regions), or pollute

fisheries (e.g., on the Papuan coast), they have to be assessed for the possibility of generating economic net losses. Similarly, however, persistence with traditional land uses that preclude or make less productive certain modern land uses involve opportunity costs.

58. In Indonesia, according to fertility data collected by Hugo, *et al.* (*The Demographic Dimension*, p. 153), Irian Jaya had the highest fertility rate of all Indonesian provinces in the late 1960s, although that came down to below the Indonesian average by 1980. In the Chittagong Hill Tracts, according to the figures collected by Dewan ("The indigenous people of the Chittagong Hill Tracts," p. 3), the indigenous population grew from about 60,000 in 1872 to 260,000 in 1951, and 440,000 in 1981. There is, however, a problem of plausibility in his figures. His figures suggest that the annual population growth rate for indigenous people in the region increased from between one and two percent in the 1951-74 period to between two and three percent for the 1974-81 period and then jumped to five percent for 1981-87, using his figure of 600,000 for 1987. The very high five percent rate does not include the additional growth required to make up for the refugee outflow and for population losses to war and genocide.

59. "Indonesia: Statement before the U.N.," pp. 91-92.

60. It is, of course, possible that the Indonesian regime's own agenda is tourism, in which case a nature reserve can, after all, constitute an extension of Indonesia's productive capacity.

61. This could even be accommodated by a particularly broad notion of productive capacity, which includes the capacity to make possible the enjoyment of nature.

62. See Peter S. Wenz, *Environmental Justice* (Albany, NY: State University of New York Press, 1988).

63. Cited in Amnesty International, *Bangladesh*, p. 7.

64. World Bank, *Indonesia*.

65. For example, in Java "some 80 percent...live in rural areas where one-third of the land is controlled by only one percent of the land owners." (Nietschmann, "Economic development by invasion," p. 3.)

66. For a fuller development of this general argument as well as the argument that the affluent countries of the world also come into this picture, see Penz, "Development refugees."

67. See Mey, *Genocide in the Chittagong Hill Tracts*, p. 102; Ahsan and Chakma, "Problems of national integration," p. 969. In India, which Bangladesh was part of until 1947, the term "backward" is an official designation for certain disadvantaged groups.

68. Cited in Burger, *Report from the Frontier*, p. 142.

69. Juwana, "Observer of the Republic of Indonesia, Statement to the fourth Session of the Working Group on Indigenous Populations of the Sub-Committee on Prevention of Discrimination and Protection of Minorities, United Nations, Geneva, 1 August, 1985"; cited in Nietschmann, "Economic development by invasion," pp. 10-11.

70. Bhikku, "Bangladesh," p. 25.

71. There is, of course, still a fundamental conflict between the efficiency logic and the inalienability of land under the tribal economic system.

72. Robert Goodland, "Tribal peoples and economic development: The human ecological dimension," in John H. Bodley, ed., *Tribal Peoples and Development Issues: A Global Overview* (Mountain View, CA: Mayfield, 1988), pp. 399-400.

73. For early work on this issue, see W.H.R. Rivers, *Essays on the Depopulation of Melanesia* (Cambridge: Cambridge University Press, 1922).

74. With respect to the lifestyle that tribal peoples lose, Marshall Sahlins ["Notes on the original affluent society," in R. Lee and I. DeVore, eds. *Man the Hunter* (New York: Aldine Publishing Co., 1968), pp. 85-89] has argued that at least hunter-gatherers have lived under conditions that represent the "original affluent society," in that they provide for considerable leisure, wants are limited by what can be carried from location to location and normally hunter-gatherers have been able to find environments that provide abundantly for those modest wants.

75. See Burger, *Report from the Frontier*, ch. 3; Goodland, "Tribal peoples," p. 404.

76. M. Colchester, "Unity and diversity: Indonesian policy towards tribal peoples," in *The Ecologist*, Vol. 16, Nos. 2/3, 1986, p. 89.

77. Mey, *Genocide in the Chittagong Hill Tracts*, p. 114; Chowdhury, "Self-determination," p. 298.

78. "Bangladesh," in *Economic and Political Weekly*, p. 2125.

79. With respect to the diversity of ethnic and national groups, there is a sharp difference between Indonesia and Bangladesh. Whereas Indonesia is generally recognized as one of the most ethnically diverse countries in the world, Bangladesh is one of the most homogenous countries, in that non-Bengalis constitute only a tiny fraction of the country's population.

80. "The political divisiveness of ethnic identity in Indonesia is such that no post-Independence government has allowed the inclusion of an ethnicity question in a census, the immense cultural, social and economic significance of ethnicity in the country notwithstanding." (Hugo, *et al.*, The demographic dimension, p. 18.)

81. Chowdhury, "Self-determination," pp. 295-96.

82. Allen Buchanan, "Toward a theory of secession, in *Ethics*, No. 101, 1991, p. 328.

83. Buchanan, "Toward a theory of secession," pp. 328-29.

84. Ahsan and Chakma, "Problems of national integration," pp. 959-60.

85. Buchanan, "Toward a theory of secession," p. 328.

86. See, for example, Goodland's ("Tribal peoples and economic development," p. 392) "four fundamental needs of tribal societies [which] relate to autonomy and participation, to conditions that will maintain their culture and their ethnic identity to the extent they desire: (a) recognition of territorial rights, (b) protection from introduced disease, (c) time to adapt to the national society, and (d) self-determination." Goodland's (pp. 397, 403) self-determination requirements are, however, expressed in more circumspect terms, given the report's intended role as World Bank policy.

87. Buchanan, "Toward a theory of secession."

88. Buchanan, "Toward a theory of secession," pp. 329-30.

89. Buchanan, "Toward a theory of secession," pp. 330-32.

90. OPM, "Indonesia, pp. 92-93.

91. Chowdhury, "Self-determination," p. 300.

92. Nietschmann, "Economic development by invasion," p. 2 (emphasis in the original).

3

Territory, Forests, and Historical Continuity: Indigenous Rights and Natural Resources

Susana B.C. Devalle

Sand, stone, soil, land, trees,
the Government takes it all...The
Government denies us our rights...
—*a Munda elder*

When "economic imperatives" are imposed on territories rich in natural resources that are, at the same time, the historical habitat of politically subordinated indigenous populations, we observe these "imperatives" often being opposed by the efforts of the indigenous populations to defend their endangered social and historical continuity. This chapter seeks to examine the issue of the rights of indigenous populations over natural resources in their territories, and the manner in which the defense of these rights has been formulated in political practice. I will particularly refer to situations in which ethnic ascription, and economic and political subordination correlate. As an illustrative case I will discuss the struggle of the *adivasi* (so called "tribal") peasantry for the forests in Jharkhand in the state of Bihar (India).

The adivasis of Jharkhand have not benefited from "development." Instead, they have been adversely affected by it. Natural resources (minerals, timber) are extracted from the Jharkhand region of Bihar and delivered to the more developed areas in the plains or exported out of India. In this context, in addition to being displaced from their lands to make room for industrial and commercial schemes, the adivasis of

Jharkhand have been practically excluded from employment in the modern sector or marginally maintained as a pool of cheap labor. Furthermore, it should be noted that "development" in Jharkhand is taking place under the threat of guns, and in the general context of violence that characterizes present-day Bihar.

I consider the struggle for the forest in Jharkhand to be an expression of a culture for resistance, developing in opposition to what I have defined as a culture of oppression which operates throughout Bihar. The oppositional grass-root politics of the adivasis of Jharkhand are directed towards aims that are not limited to the economic level alone. These aims attempt instead to embrace realms of the social reality which the mechanisms of formal politics and the state have subordinated to economic imperatives and to the "higher" goals of a sectional "national interest." The struggle for the forests in Jharkhand forms part of a wider political discourse in which life—not just "survival"—is defined in its widest sense: as the defense of the adivasis' physical, historical, social, and cultural continuity. At the same time, this defense expresses a strong indictment of the dominant discourse on "progress."

While addressing these problems, I will refer to the meaning of the loss of territory (and therefore, the loss of rights over its natural resources) has for indigenous peoples. The so-called "felling craze" will be discussed as an instance of the struggle for the forests in Jharkhand. An examination of the particularities of the Gua incident in 1980, will disclose the operation of the dialectics of oppression/resistance around the question of the use and exploitation of the forest of Jharkhand.

Territory and Historical Identity

A people's consciousness of long permanence in a given territory with specific social and historical meanings, translated into a sense of belonging to a place, is usually one of the terrains in which collective identity is formulated and reformulated over the course of time.

For indigenous peoples, territory is a space that is not merely "used" but fundamentally "lived." History, culture, and territory merge in the definition of collective ("ethnic") identity. Thus, territory acquires a double meaning. Land is at the same time a means of production and the ground where the social and cultural world is ordered.

The processes of total or partial dispossession of the indigenous populations of their rights over their lands/territories and the natural resources these territories contain have followed a common pattern, marked by specific historical circumstances: the colonial expansion, the development of unequal socio-economic relations that permit the exploitation of labor, land, and natural resources, the nature of the modern state, and the models for economic development adopted by the states of the so-called Third World. Regarding land/territory, these processes entail a dispossession that goes beyond the mere geographical-ecological dimension.

The process of dispossession has been aided by legitimating ideologies that, phrased in racial or ethnic terms, support existing patterns of political domination and unequal socio-economic relationships. In the case of India, the "tribalist" perception, while constructing categories that refer only to ideal models, has provided elements for the formulation of a very concrete and functional ideology. This ideology serves to support and reproduce capitalist relations of production and patterns of power relationships, and to justify the expansion of cultural hegemony.[1] In India, the official category Scheduled or Backward Tribes (under which the indigenous populations are catalogued), artificially determines a "special" social sector with "special" problems (phrased as weakness and backwardness), deserving a "special" treatment and in need of guidance towards "development." In consequence, the objective situation of the populations thus catalogued is ignored, and the structure of inequality is validated by attributing its causes to deficiencies assumably inherent to these subordinate sectors.

Indigenous peoples have opposed the processes of de-historization and deculturation—that is, processes of total dispossession—with their continuous efforts to maintain their socio-cultural specificity. The defense of the territory is one of the axes around which these efforts have been centered.

Not surprisingly, the defense of land/territory appears as a key issue in the political discourses of indigenous movements world-wide. These movements devote a great part of their efforts to regain control over indigenous historical habitats. In Jharkhand, these efforts are translated into demands for regional autonomy.

Jharkhand: The "Ruhr" of India

Greater Jharkhand is conceived as a cultural region extending over Chotanagpur and Santal Parganas in Bihar, parts of West Bengal, Orissa, and Madhya Pradesh. This is an area rich in mineral and forest resources. The largest adivasi population of India is concentrated in this region (ten million persons registered as Scheduled Tribes).

The adivasis of Jharkhand are basically peasants: eighty-nine percent of them live by agriculture. Forests (29.2 percent of the land area in Jharkhand) supplement the agrarian economy. The expansion of industries in Bihar since the 1950s and the development of commercial exploitation of the forests have accelerated the process of land alienation, limited alternative sources of subsistence, and undermined the adivasi peasant economy.

Jharkhand has been the setting of a sustained agrarian-based tradition of protest with ethnic overtones since the end of the eighteenth century. In present-day Jharkhand this tradition continues not only in political movements but also in everyday life in a culture for resistance, the counterpart of a culture of oppression, now a pervading feature in Bihar.

The movement the Jharkhand Mukti Morcha (JMM) began to lead in the Santal belt in 1972 aimed at the recovery of alienated lands. Together with the forest movement in Singhbhum, it formed part of an alternative project for Jharkhand. Until the early 1980s, the JMM presented an alternative both to the state's "development" project and to the reformist project spoused by the adivasi urban petty bourgeoisie (represented, among other organizations, by the Jharkhand Party).

The Dying Forests

The adivasis used to reclaim land from the forest for agricultural purposes, and had customary rights over the collection of forest products. The forest policy of independent India was formulated in 1952, stating that village communities would not be permitted to use the forests "at the cost of national interests."[2] The rights of the adivasis to cultivate foodcrops and to collect forest products were transformed into "concessions." The effects of this policy have been considered disastrous.[3]

Plans for the commercial exploitation of the forests—and not an assumed ecological damage caused by the way they have been traditionally used (as stated by the National Commission of Agriculture)—are the reason for the growing exclusion of the adivasis from the use of forest areas.

The introduction by the Forest Development Corporation of teak plantations in Jharkhand, with purely commercial goals, have resulted in continuing and violent tension in the area. Not only is teak useless for the Jharkhandis (unlike sal [Shorea robusta], which is being replaced), but it also tends to adversely affect the soil, so that other crops cannot grow at its side. Other trees, and particularly sal trees, also have important non-economic meanings for the adivasis (for instance, the sacred sal groves.

Massive deforestation has brought the total area under forests in Jharkhand from thirty-three percent in 1947 to ten percent in 1980. Droughts and soil erosion, both with disastrous consequences for agriculture, have increased with deforestation. Peasants have been displaced from lands in the forests which were suitable for cultivation the moment the government decided to introduce commercial trees.

In addition, some areas of Jharkhand have become suppliers of specific kinds of labor, in a way perhaps similar to the "labor reserves" categorized by Samir Amin for Africa,[4] but without an institutionalized segregation. The "forest villages," called significantly "labor camps" by the Forest Department, have functioned as small reserve-area units with a practically captive labor force.

The Dialectics of Cultural Struggle

The achievement of hegemony in Bihar has been sought through the operation of a culture of oppression. At the same time, counter-hegemonic practices have become integrated into an oppositional culture for resistance.

Culture is conceived here as a way of life molded by social and economic forces, entailing a whole social order which involves a set of signifying practices—the languages in which a cosmovision supporting a social order is expressed—and a way of feeling—the subjective experience of the social world that allows one to link, to use Samuel's words, "the individual moment with the longue duree."[5]

Implicit in the culture of oppression is the conception of violence as a "right" of the powerful, while those who are the targets of this violence are viewed as the natural, almost de-humanized "objects" of this violence. In the Gua incident of 1980, related to the forest movement in Singhbhum, to be discussed later, we can observe the dynamics of a "theatrical style of dominance" and its transformation into practices of terror.

A culture for resistance, on the other hand, is active not only in open revolt, but is continuously operative in social consciousness and in actions of opposition developed by the subaltern sectors in everyday life. This culture supports a social strategy for survival, self-assertion, and, ultimately, physical, historical, and cultural reproduction.

The "Felling Craze"

The ecological non-violent Chipko movement (which started in 1973 among Himalayan low caste peasants) is possibly the best known of the movements of this kind in India. The symbolism of hugging the trees to save them that the Chipko movement conveys contrasts sharply with the so-called "felling craze" in Jharkhand and what has been called "a periodical sort of frenzy" of tree-felling by adivasis in Bastar.[6] The historical and social contexts in which these forms of protest have emerged are different. In addition, the last two cases present a long history of dispossession and violence by the powerful.

The opposition to the replacement of the sal forest by teak was met with ruthless police repression, launched even before any confrontation had taken place, as a "preventive measure." This happened, for instance, as a response to rumors that teak trees would be felled, and in Serengda village as a response to the villagers' symbolic cutting of some teak saplings.[7] As an adivasi put it: "Teak is killing us." This is not just a metaphor. These repressive tactics can be seen as exercises in the theater of power and as manifestations of the culture of oppression in Bihar.

Thus started, the so-called "felling craze," a consistent campaign in defense of land and the traditional forest, was considered by the authorities and much of the press as a display of madness. A local political activist explained the situation in the following way:

> The Government wanted to remove sal and mahua to plant commercial teak. Nobody then could use the forest. We [the adivasis] cannot use teak, so we cut the trees. We feel that the traditional forest is being destroyed...There is nothing we can do and then it is better to make the land cultivable [by cutting the useless teak].[8]

Meanwhile, the JMM leader, Shibu Soren, curtly stated: "Adivasis will stop felling trees when their needs are met."

The authorities were keen to suppress the movement by force. Why this reaction out of proportion? On the one hand, in 1980 it was feared that the protesters' threats to stop dispatches of coal, iron-ore, bamboo, timber, and other mineral and forest products out of Jharkhand could materialize. On the other hand, the felling of trees has been directly related to the demands for land restoration and for a Jharkhand state put forward by the JMM. The JMM was backing the jungle kato movement in Singhbhum with the aim of claiming land for cultivation in the forests tracts. The repression of the forest movement in Jharkhand's Singhbhum district is considered to have been part of a "crush the tribal movement" offensive in Bihar in the 1980s.[9] The people's confrontation with the Government agencies culminated in September 1980 in the Gua massacre.

Gua: Power Translated into the Practice of Terror

On 8 September 1980, a procession of adivasis, wishing to submit a memoranda to the forest officials, was allowed to march into Gua and hold a public meeting organized by the Jharkhand Mukti Morcha (JMM). When the police arrested two of the leaders and attacked the multitude, the adivasis answered back by shooting arrows, killing two policemen. The police fired, killing some and injuring others. The counterpoint between power and subalternity, was broken at this point. The injured were taken by their companions to the hospital where police forces were already waiting. The police lined them up and shot them. Thus, the Gua confrontation concluded with an instance of the practice of terror: the killing of the innocents at the doors of the hospital performed as exemplary punishment.

In symbolic terms, the punishment inflicted on the innocents stresses before the people the unlimited reach of the violence inspired by the powerful's wrath. Terror comes to be seen as a force that nobody can contain. It is perceived by its "objects" as unending and beyond human control. I recorded the following comment shortly after the Gua killings: "We are killed and robbed. It happened yesterday, [it happens] today, [it will happen] tomorrow."[10]

After the Gua incident, adivasis set fire to some thirty-five bungalows and rest houses in the forests, these seen as the material symbols of authority in invaded adivasi territories. At the same time, no judicial inquiry was made into the Gua firings.

A Political Ecology?

The urgent need to assure physical and social continuity is a common denominator in all modes of popular resistance. In the case of subordinated indigenous populations, to survive means to maintain their continuity as a people, to emphasize their historicity, to affirm the specificity of their particular socio-cultural style. In a word, the indigenous peoples' effort at survival is a struggle to be, to exist in historical and social terms. In Jharkhand, from different political and class angles, the adivasis acknowledge that their existence is at risk.

In the course of this struggle to exist in holistic terms, cultures for resistance are forged. There is, however, no inherently or perpetually rebellious culture. Culture becomes rebellious. As the Jharkhand case shows, the situation of subordinated indigenous populations facing the imposition of exogenous "economic imperatives" backed by the practices of a culture of oppression, calls for the development of a rebellious culture. Implicit in this cultural struggle is a formulation of alternative social projects which consider basic human rights and needs.

In this light, the explicit defense indigenous peoples make of their territories and the natural resources these contain, appears to be one of the "texts" in which the discourse of resistance is expressed. This discourse opposes the discourse of power that used a sectorial concept of "progress" as an instrument for further domination. In T. Nairn's words, "on the periphery...[people] learnt quickly that Progress in the abstract meant domination in the concrete."[11]

The implementation of "progress" in Jharkhand has produced a situation marked by the poverty of development. Is this "progress" worth the violence and death to which it has subjected the Jharkhand people? Is it only coincidental that "development" should have been equated by some authors with war? For instance, for Naim: "Uneven development is a politely academic way of saying 'war',"[12] and for Galtung: "The more development, the less peace."[13] The traditional opposition North-South (or rich-poor, in itself an unequal opposition), continues to rule the views on development from a dominant position, views that have also permeated the optics of the planners in the so-called Third World. A new perspective has been emerging that concentrates the attention of the "South" on the gaze of "the poor" on the "South" itself, supporting a qualitatively different view on societal development that takes into account the needs and aims of each particular society.

Much is now discussed on sustainable development. The initiatives for this kind of development, however, should come from the same societies where this development is to take place, and be enriched by a dialogue with other societies sharing similar historical, social, and political experiences (the South-South dialogue). To this, another aspect should be taken into consideration: the proposals for ethnodevelopment coming from indigenous peoples.[14]

In this context and with reference to natural resources, in the case of forests it is not sufficient just to develop the field of "social forestry" (where often trees seem to be at center stage while people's concerns are submerged). In the case of indigenous peoples such as the adivasis in India, the issue has always been "beyond mere ecology." In view of the way in which the defense of territory, land, and natural resources has been posed by indigenous peoples, one could venture to speak in terms of a political ecology, a perspective on ecology that takes into account global economic processes, and views ecology not as an autonomous subject but closely linked to processes of social and cultural continuity, with meanings to be found in the deep historical field.

Notes

1. For Africa, see A. Mafeje, "The ideology of tribalism," in *The Journal of Modern African Studies*, Vol. 9, No. 2, 1971, pp. 253-261; H. Wolpe, "The theory of internal colonialism: The South African case," in I. Oxaal, T. Barnett and D. Booth, eds., *Beyond the Sociology of Development* (London: Routledge and Kegan Paul, 1971), pp. 229-252. For India, see Susana B.C. Devalle, *La Palabra de la Tierra:*

Protesta Campesina en India, Siglo XIX (Mexico: El Colegio de Mexico, 1977) and *The Dialectics of Cultural Struggle: Discourses of Ethnicity, Jharkhand* (Beverly Hills, Sage, 1992).

2. K.P. Kannan, "Forestry: Forests for industry's profits," in *Economic and Political Weekly*, Vol. 17, No. 23, 1982, p. 936; also see S. Kulkarni, "Towards a social forest policy," in *Economic and Political Weekly*, Vol. 18, 5 February, 1983, p. 192.

3. N. Sengupta, ed., *Fourth World Dynamics: Jharkhand* (Delhi: Authors Guild, 1982), p. 17.

4. Samir Amin, *El Capitalismo Periferico* (Mexico: Nuestro Tiempo, 1974), p. 172.

5. R. Samuel, ed., *People's History and Socialist Theory* (London: Routledge and Kegan Paul, 1981), p. xxxii.

6. A. Sharma, "Felling in Bastar forests unabated," in *Times of India*, 21 May, 1990.

7. See, "Repression in Singhbhum," in *PUCL Report*, March, 1979; "Singhbhum. Exploitation, protest and repression," in *Economic and Political Weekly*, Vol. 14, No. 22, 1979, pp. 940-943.

8. Interview.

9. G. Pardesi, "Gua incident: Operation annihilation?," in *Mainstream*, Vol. 19, No. 11, 1980, p. 7.

10. Comment recorded in the Chota Nagpur/Singhbhum area.

11. T. Nairn, "The modern Janus," in *New Left Review*, No. 94,.1973, p. 10.

12. Nairn, "The modern Janus," p. 15.

13. J. Galtung, *Peace and Development in the Pacific Hemisphere* (Honolulu: University of Hawaii, Institute for Peace, 1989), p. 3.

14. Rodolfo Stavenhagen, "Introduccion," in R. Stavenhagen and Nolasco, eds., *Politica Cultural para un Pals Multietnico* (Mexico: SEP and El Colegio de Mexico, 1989), pp. 7-18.

4

Small-Scale Mining and the Environment in Southeast Asia

Michael C. Howard

Small-scale mining has received a great deal of attention in Southeast Asia in recent years, especially the "gold rushes" of the 1980s in the Philippines and Indonesia. Small-scale miners have been at the center of much of the debate surrounding how the mining industry should best be developed in relation to its social, economic, and environmental impact. While media attention has focused on the dramatic, such as the deaths of hundreds of miners in the southern Philippines as a result of cave-ins, small-scale mining have also been debated and discussed by academics, government officials, lobbyists for mining companies, politicians, native rights advocates, and environmentalists.

Sentiment has been pretty well divided. While government officials and mining companies have been highly critical of the activities of small-scale miners, liberal academics and native rights advocates, in particular, have been supportive. Environmentalists outside of government, for the most part, have also been supportive of small-scale mining as a preferable alternative to large-scale mining. Those within governments responsible for environmental protection, however, have been critical of small-scale miners. In the discussion that follows, after a review of the development of the mining industry in the region, I will discuss and analyze the debate surrounding small-scale mining with particular reference to environmental questions. Such environmental issues will be viewed within a political and economic context and as part of a discussion about developmental options.

The Mining Industry in Southeast Asia Today

The importance of mining varies among the states of Southeast Asia. It is of most importance in the Philippines, Malaysia, and Indonesia, which have large and well-established mining industries. It is of somewhat less importance in Thailand and Burma and relatively insignificant in Vietnam, Laos, and Cambodia.

Although industrialization has led to increased local demand for metals, the mining industries of most Southeast Asian countries historically have produced primarily for export. For this reason, they are particularly susceptible to fluctuations in the international market. Thus, declining prices in the mid-1980s, with world commodity prices hitting a seven-year low in 1986, caused widespread problems for the industry. To some extent, however, these were offset by technological developments in certain sectors (such as gold) and growing internal demand in some economies (such as Thailand's). Gradually improving prices toward the end of the 1980s helped the industry as a whole to recover, but subsequently prices have again declined for most metals and prospects for the immediate future are not encouraging. Additional pressure on the industry has resulted from increased environmental concerns.

The Philippines is a major producer of minerals. It ranks eighth in world gold production and tenth in copper production. Mining takes place in many parts of the Philippines, including the mountains of northern Luzon (the Cordillera), the islands of Cebu and Marinduque, and northern and eastern Mindanao. Among the largest mining companies are Atlas Consolidated, Marinduque Mining, Benguet Corporation, Philex Mining, and Lepanto Consolidated.[1] All of these companies are closely associated with powerful Filipino interests, as well as having ties to Japanese, American, and other foreign interests. Political instability and poor international market conditions hurt the country's mining industry during the early and mid-1980s. Internal political and economic problems meant that the country largely missed out on the gold boom experienced by neighboring countries in the 1980s. Although the Philippines is believed to have significant epithermal deposits, such as are being exploited in Papua New Guinea and elsewhere in the South Pacific,[2] the richest mineralization for new gold production in the Philippines is to be found in eastern Mindanao, where concern over tangling with guerrillas and small-scale miners frightened away would-be investors.

The mining industry in the Philippines recovered during the late 1980s as a result of increased internal political and economic stability and improved metal prices, especially for copper (which accounts for almost sixty percent of the country's mineral exports by value). Production of copper and gold as well as of other metals, such as nickel, chromite, manganese, and iron ore increased at the end of the decade, and the economic performance of most mining companies improved. The gold industry faced a temporary setback as a result of the 16 July 1990 Baguio earthquake. Earthquake damage aside, there was some optimism in the mining industry in 1990-91. Benguet invested substantially in the new open-pit Grand Antamok Gold Project and began construction of a new "carbon-in-leach" mill and expanded its Baco and Kelly underground mines. Lepanto (sixty percent) and CRA of Australia (forty percent) formed the Far Southeast Gold Resources joint venture in 1991, and announced plans to invest US$300 million to develop an open-pit mine in the Cordillera capable of producing 250,000 ounces of gold per year by 1995. In addition, a new gold mine, Nalesbitan, began operation in Camarines Norte in early 1990

Mining investment has also picked up in Mindanao. Banahaw Mining Corporation, a subsidiary of Muswellbrook Energy and Minerals of Australia (in which Kerry Packer has an interest), started to operate the first new mine to open in the Philippines in a decade on Mindanao. The operation got off to a rough start, however, when it was attacked by NPA guerrillas in January 1989. The attack left five employees dead and caused some damage. In response, the mine was provided with 260 armed guards by the army.[3] The Banahaw mine has been watched carefully by other Australian mining interests, who remain wary of guerrillas and small-scale miners. During the same year, Benguet Corporation commenced drilling at its Zamboanga Gold Prospect, the site of a potential open-pit gold mine.

By 1992, however, it was clear that the country's mining industry was still in a state of decline. The overall value of mineral production in the Philippines declined by fourteen percent in 1992, with copper production at half its early 1980s level and gold production at its lowest level since 1980. In an effort to overcome the lack of exploration and investment capital in the industry, the Aquino government had enacted a new mining code allowing complete foreign ownership of mining ventures, but the code appeared to be contradicted by the new constitution, which required majority Filipino interest. Under the mining code the Ramos government has sought to promote financial and technical assistance agreements which allowing foreign mining interests to undertake exploration on their own, and to reduce their holding in any subsequent venture to forty percent over a period of up to twenty years and after recovering their initial investment. The first of these agreements was signed in early 1993 with Arimco of Australia to explore for copper and gold in Nueva Vizcaya.[4]

Malaysia has long been known as a major tin producer, but in recent years tin mining in Malaysia has been an industry in decline. In 1989 tin exports were valued at just over US$338 million, compared to more than

US$10.5 billion for the country's other major exports. Following its collapse in 1985, the international tin market slowly regained ground following the formation of the Association of Tin Producing Countries in 1987. Malaysia, however, found itself having to share its spot as a top producer with Indonesia and Brazil. Continued low prices and sluggish demand, in contrast to other commodities produced by Malaysia (such as rubber and palm oil), has served to shift interest away from tin mining. The industry also faces problems from the depletion of reserves. The technology is available to allow dredges to recover tin from previously inaccessible depths, but many tin mining companies are more interested in diversifying into property, gaming, and recreation, and converting land to other uses.[5] The government, for its part, while issuing occasional words of encouragement to the industry, has withdrawn many of its incentives and put more land aside for industrial development.

Indonesia's mining industry has received far less attention than its petroleum and forest industries. While mining represents only 1.4 percent of GDP in Indonesia, especially in Kalimantan, Sulawesi, and Irian Jaya, mining is of considerable importance. The downturn in the international market in the early 1980s, as in the Philippines, depressed the mining industry in Indonesia. The subsequent upturn in the market in the late 1980s, however, led to significant growth in the mining sector, especially in gold and coal exploration. Most of the new activity has been carried out by Australian companies.[6] By 1988, some one hundred companies had taken out mining concessions, mostly for gold exploration. Since then, the number of companies continuing to be active has declined.

The one mine in Indonesia that has received considerable critical international attention is the large Freeport copper mine in Irian Jaya, which has been embroiled in the struggle over the political status of Irian Jaya. The high grade of its copper ore and its relatively large gold deposits kept the Freeport mine profitable throughout the 1980s. As metal prices improved in the late 1980s, company profits, along with the amount it paid in taxes and royalties, increased substantially. In late 1989, Freeport-McMoran announced a major expansion program for the Irian Jaya mine.[7] The announcement followed a new find at nearby Grasberg Mountain that doubled proven copper reserves and substantially increased proven gold reserves. Production of the mine, which had fallen in recent years, increased in the early 1990s as a result of the new discoveries. With its mill output exceeding 60,000 tons in 1993 and expecting to reach 90,000 tons in 1996, existing reserves in 1993 were anticipated to last until 2016.[8] An Australian geological survey in 1989 resulted in Freeport being given exploration rights over an additional 25,000 square kilometers along the Carsteng mountain range. The geological survey also indicated significant deposits of gold near Enarotali, not too far from the Freeport mine.

What has received far less attention is Indonesia's most recent "gold rush" which began to take shape in 1986. The rush began in earnest the following year, coinciding with a substantial increase in gold exploration in Papua New Guinea and elsewhere in the southwest Pacific.[9] In Indonesia, commercial exploration focused on Kalimantan, where timber operations had opened up many previously inaccessible areas by building roads and airstrips, and northern Sulawesi. Foreign mining companies were attracted by the relative strength of the gold market—as noted by one mining executive, "Gold is the only commodity making money at the moment."[10] An additional incentive was the degree of political stability in comparison with, for example, Papua New Guinea and the Philippines. Australian, American, and British interests were involved in gold exploration in Kalimantan, primarily around Mandor, some eighty kilometers north of Pontiak, and near Mount Muro, in central Kalimantan. Commencement of commercial mining got off to a slow start and results of much of the initial exploration has been disappointing. By 1991, it appeared as if mines would be developed by three companies: Duval, Billiton, and CRA.

Kalimantan has also become the focus of intense activity in coal production. Industry sources estimate that coal exports from Kalimantan will reach twenty million tons by 1992, increasing to thirty-eight million tons by 2000—representing one-third of the world steam coal market.[11] Destined for markets in Japan, Korea, and Europe, the coal will come from South Kalimantan and East Kalimantan, and be shipped from a port to be developed on Pulau Laut. The largest mining operation is a joint venture between BP Coal and CRA, under the name of Kaltim Prima Coal (KPC), which was formed in 1988 to develop a large opencast mine near Samarinda in East Kalimantan. Other foreign companies involved in developing coal projects in Kalimantan include BHP-Utah of Australia, New Hope Colleries of Australia, and Utah Indonesia.

Interest in diamond mining in Kalimantan also has grown in recent years. By the late 1980s, small-scale mining (primarily by Javanese) centered along the shores of Riam Kanan involved an estimated 30,000 people. After initial setbacks following the October 1987 stock market crash, Perth-based Indonesian Diamond Corporation announced that money was being raised to start Kalimantan's first large commercial diamond mining project. It was to be a dredging operation intended to produce around 120,000 carats of gem diamonds per year.

Thailand's mining industry in the past was dominated by tin, but the poor state of the international tin market and the growth of other sectors of Thailand's economy since the mid-1980s has reduced the importance of tin mining. In recent years mineral production has been increasingly oriented toward the domestic market rather than the export market. Zinc, for example, ceased to be exported in 1991, as the country's entire production of zinc was consumed domestically.[12] This shift in orientation has resulted in greater emphasis being placed on lower priced industrial minerals. Thailand has also been a major producer and exporter of gemstones, with the value of gem exports growing rapidly during the late 1980s. By 1990, gems and jewellery ranked as the country's third largest export item, and Thailand ranked second worldwide after India.[13] While in the past most mining was done by small-sized and medium-sized operations, large companies now dominate much of the industry. Depletion of gem deposits close to the surface, for example, has meant that most gem mining is now carried out by large-scale miners employing mechanized and expensive techniques. Small-scale gem miners, for their part, came to search for stones amidst the fighting in western Cambodia.

The development of mining in Burma has been hampered by the country's on-going civil war and much of what is produced is smuggled out of the country. Decades of warfare ensured that Vietnam, Laos, and Cambodia were unable to develop their mining industries. Cambodia's gem fields near the Thai border featured prominently in fighting between the Khmer Rouge and government troops, with revenue from the fields becoming an important source of revenue for the Khmer Rouge.[14] Since the end of hostilities in the 1970s, the governments of Laos and Vietnam have sought to develop their mining industries, but, so far, with only minor success.[15]

Historical Background

Southeast Asia has a relatively long history of small-scale mining. The origins of metallurgy in the region appear to lie with the Dong-son civilization that spread along the northern Vietnamese coastline from about B.C. 300, and exerted influence throughout much of Southeast Asia. Dong-son civilization produced a variety of bronze implements, acquiring a knowledge of bronze-making through extensive trading links. Precisely how and where it acquired the knowledge is unknown. Archaeological evidence for the subsequent spread and development of mining and metal-working is scattered and sparse and more in known about the objects produced than where the material to make them came from.[16] Small-scale alluvial mining of gold appears to have been widespread, with various accounts by Chinese, Arabs, and Indians referring to mining and metals in the region dating from around the fifth century A.D. Somewhat more intensive mining techniques appear to have emerged relatively late and only in a few locales.[17] The technology employed and the scale of mining meant that the environmental impact was negligible.

Fresh from their success in the New World, the Spanish had great expectations of finding large quantities of gold in the Philippines. What they found instead were small alluvial deposits and sufficient resistance from local inhabitants to lead them, effectively, to abandon any hope of finding important mineral riches. Most of the gold that was eventually to be mined in the Philippines was beyond the resources and technology available at the time, and mining on a significant scale was not to be attempted in the Philippines by Europeans again until the 1890s.

Mining activity in Southeast Asia began to increase in the in the late eighteenth century as a result of Chinese and, to a lesser extent, European interest.[18] Starting in the late eighteenth century, the Chinese became increasingly involved in tin mining along a line of deposits running from southern Thailand to a group of islands off the east coast of Sumatra and in mining gold in northwestern Borneo. They were given mining concessions by local Malay rulers, desirous of the potential income from rents and taxes, who considered the Chinese to be better miners than their own subjects. In the Malay peninsula, the Chinese continued to rely primarily on Malay laborers until the mid-nineteenth century, while elsewhere mine-work came to be done by Chinese. The Chinese were organized into associations that operated with almost complete independence from the Malay rulers. The mining was labor-intensive and the technology tended to be fairly simple. Gold production in Borneo, for example, relied primarily on sluices and pans. The number of people involved was relatively large in contrast to the local population. Raffles estimated that there were 32,000 Chinese working the northwestern Borneo goldfields by 1812.[19] The Chinese influx into the Malay peninsula came a little later as a result of the spread of British rule. In the 1820s, there were only a couple of thousand Chinese tin miners in the peninsula, but in the 1860s and 1870s, as important new tin deposits were discovered, tens of thousands of them began to arrive.

The real increase in the scale and intensity of mining in the region, however, was to come in the mid-nineteenth century as a result of growing demand for metals in Europe and North America, especially for tin, and related technological innovations.[20] Between 1870 and 1900 world demand for tin increased by three times. Dutch and British commercial and colonial interests invested in tin mining in Indonesia and Malaya respectively, greatly expanding the industry. The Europeans dominated capital-intensive mining, while the Chinese continued to operate smaller mines and to provide labor for the European-owned mines. In Indonesia, large tin mines were developed on the islands of Billiton, Bangka, and Singkep, located south of Singapore and east of Sumatra. Billiton Tin Company was founded in 1852 by private interests and mining began in 1861. The Billiton mine became the largest and most profitable mining concern in the colony.

The scale of tin mining in the Federated States of Malaya grew rapidly following the discovery of major deposits in Perak in 1848. By 1898, Malaya was the world's leading producer of tin. At first the Chinese continued to dominate the industry. New technology, in the form of a steam engine and centrifugal pump, was introduced in 1877, and the first European tin-mining company was formed in 1884. European capital did not assume a major place in the industry, however, until after the onset of the tin boom of 1898, and, especially, with the introduction of the bucket dredge in 1912. For their part, the Chinese continued small-scale alluvial tin mining and still provided most of the labor in the larger operations. In fact, the tin boom ushered in a huge influx of Chinese immigrants (100,000 in 1899 and 1900 alone).

Small commercial gold-mining operations were scattered throughout island Southeast Asia. Larger operations were to be found in Sarawak and the northern Philippines. By the middle of the nineteenth century the amount of gold being produced in southwestern Borneo was in decline and conflicts erupted among the Chinese miners and between them and the Brooke government and the Dutch. Gold production during the remainder of the century in Borneo was limited. In 1898, however, the Second Rajah Brooke established the Borneo Company which introduced more intensive gold mining methods in Sarawak. The company continued operating until 1921, after which local Chinese obtained leases and continued to work the depleted goldfields on a small scale.

In the Philippines, the Spanish had shown little inclination to promote mining after their failures in the early seventeenth century. Activity picked up in the 1890s, when European interests undertook limited exploration for gold and began to exploit alluvial deposits primarily in the Cordillera, but the unsettled conditions caused by the revolt against Spain and American intervention halted activities before any real progress was made. The American administration provided a far better institutional infrastructure for mining than had the Spanish, and American prospectors with new technology ushered in a boom in mining exploration and development in the Cordillera. Capital was the biggest problem facing most mining entrepreneurs. The most successful operation was Benguet Consolidated Mining Company. Founded in 1903 to exploit gold deposits in the Antamok area, near Baguio, after some initial difficulties, the company came to dominate mining in the Cordillera. Gold-mining underwent a boom in the Philippines between 1933 and 1936, and several new mines were established in the Cordillera and elsewhere in the country. Capital for mining came primarily from the United States, Britain, and the Philippines itself.

Coal was the only other substance to be mined at this time in significant quantities. Commercial coal mining began in Indonesia in 1849. Three of the more important mines were those of Bukit Asen (Bukit Asam) and Ombilin in southern and western Sumatra and Pulau Laut, off the southeastern corner of Kalimantan. Coal production was relatively small until after the First World War. By the late 1930s, some two million tons were being produced annually. Much of the labor in the coal industry was provided by Javanese, including a number of convicts.

The contrast between commercial mining of the late nineteenth and early twentieth centuries and mining that had gone on before is considerable. The major jump that occurred in the scale and intensity of mining meant that production soared and, along with it, so did the wealth produced from mining. In keeping with the past, however, most of what was mined was still carried away, along with most of the wealth, with local elites skimming off a little for themselves. In fairness, it should be noted that at least some of the revenue from mining went into colonial government coffers to pay for local infrastructure. The nature of mine-work was transformed dramatically. Taking two extremes, the convicts working in the coal mines of Pulau Laut were a far cry from the Dayak sifting through river deposits for alluvial gold.

The environmental impact of mining also underwent a major change. The holes were far bigger, the piles of waste much larger, and there were new polluting chemicals with which to contend. Such environmental problems, however, received little comment at the time. The crucial questions for colonial authorities and local elites concerned who should be allowed to mine, efficiency of production in terms of maximization of output, and how to divide the revenues. Seeking to order what was perceived to be a chaotic situation, governments throughout the region enacted various sorts of mining acts, regulations, and ordinances. Devoted largely to establishing mining rights and levies, these rules paid almost no attention to environmental questions. Where they did touch on such matters, it was primarily in the form of setting safety standards for mine-workers and creating the means to enforce them.[21]

Small-scale mining activities did not disappear in the face of more intensive, large-scale undertakings. Small-scale miners continued to work areas that were uncommercial for the larger miners. They also worked around the margins of the large-scale mines, and occasionally the two sectors came into conflict. Where there were conflicts, the authorities almost always sided with the large mines. Mining regulations commonly sought to curtail the activities of small miners. The primary reason for this government attitudes was financial—large mines generated revenue for the governments and small mines, for the most part, did not.

The Second World War put a halt to most mining activity in Southeast Asia and many mines suffered considerable damage.[22] During the immediate postwar period efforts were made to rehabilitate the mining industry, but unsettled conditions in many parts of Southeast Asia and poor conditions in the international market hampered these efforts, and, in most countries the industry was unable to regain its prewar level of production. In Malaya, for example, tin production in 1947 was 27,026 tons, compared with 44,627 tons in 1939, and only about half the prewar number of mines were in operation.[23] Mining resumed in Indonesia after the war, but political and economic instability resulted in low levels of production. Under the Sukarno government most mining was carried out by inefficient and corrupt state institutions and there was little private investment in mining.[24] The mining industry in the Philippines grew slowly in the early postwar period. Faced with poor market conditions and rising costs, the gold industry was particularly hard pressed. Benguet's strategy for overcoming these problems was to diversify into timber, steel fabrication, and other areas.

The mining industry began to grow once again in the 1960s. The initial impetus came largely from steps taken by national governments. The industry then received a boost in the late 1960s and early 1970s as metal prices and demand increased. The government of the Philippines considered the gold industry and important source of foreign exchange.[25] To help support the troubled industry, it passed the Gold Assistance Act in 1961, establishing a subsidized price for gold, which was to be sold to the government. The mining industry also was provided a lift as a result of the commencement of copper mining on Cebu by Atlas Consolidated Mining (fifty-one percent owned by Phelps Dodge and fourteen percent by A. Soriano y Cia. of the Philippines) in 1955. Further support for the industry came from the declaration of martial law in 1972 and the subsequent efforts by technocrats under the Marcos regime favoring large-scale development and foreign investment. Despite market fluctuations during the 1970s, overall, the mining industry in the Philippines grew during the decade and, by 1980, mineral and metal exports (primarily copper, gold, and nickel) accounted for twenty-one percent of merchandise exports. Growth in base metal mining reflected increased Japanese involvement in the mining industry (which had previously been almost entirely controlled by Americans and Filipinos) in terms of investment capital and as a market.

The situation of the mining industry in Indonesia changed considerably under Suharto's New Order government, which sought to encourage mining as part of its general push for development. In particular, under the Foreign Investment Law of 1967, the government tried to entice foreign mining interests to carry out mineral exploration and develop mines.[26] The so-called "first generation" work contracts provided generous concessions for foreign companies. The first, and only, mining company to take advantage of the new law was Freeport Indonesia, owned by New

Orleans-based Freeport Sulphur.[27] Freeport wanted to develop one of the world's largest copper deposits, which was located in Irian Jaya. Development of the isolated site required a massive capital investment. Nevertheless, within three years of commencing production in early 1973 Freeport had earned three times its original equity investment. This very profitable mine produced an average of 188,000 tons of copper concentrate per year between 1973 and 1980, with production peaking at 229,000 tons in 1982.

The Indonesian government introduced new, less generous, rules, known as "second generation" work contracts, in 1968. The first and largest second generation contract was awarded to International Nickel of Canada (Inco), which was pursuing a strategy of international diversification and was attracted by the proximity of Indonesia to the Japanese market. Inco incorporated in Indonesia as P.T. Inco in 1968 to carry out nickel exploration in the Soroako area near Lake Matano in central Sulawesi. Inco began exploration in 1969. The poor state of the nickel market delayed development of the project, which also required a very large capital investment. When completed in 1978, the project included a nickel smelting plant with the capacity to produce one hundred million pounds of nickel matter a year, making it the world's largest. The first nickel matter was produced in 1977, with full production beginning at the end of 1978. There were initial problems with new technology and later the operation was to be plagued by the poor state of the international nickel market. Indonesia's nickel exports rose ten-fold from 1966 to 1980. The market was dominated by the Japanese and prices fluctuated widely throughout the 1970s and early 1980s.

When metal prices declined in the early 1980s, the Southeast Asian mining industry faced a severe crisis. After copper fell to sixty cents a pound in 1982, most copper mining companies in the Philippines found their operations unprofitable. The government intervened to subsidize the price at seventy-five cents a pound, but this did little to help many of the companies, especially those associated with Marcos cronies, which were heavily in debt. The industry was also confronted with labor unrest and with something new, growing concern about the environmental impact of mining, especially its impact on agricultural lands in the neighboring lowlands. The turmoil surrounding the fall of Marcos in the mid-1980s served to make matters even worse. Because of Benguet Corporation's association with Marcos's brother-in-law, Benjamin Romualdez, the Presidential Commission on Good Government assumed effective control of the company in 1986. In 1984, only five of the country's twenty-four main mining companies were showing a profit. Benguet recorded its first postwar loss in 1985, before returning to profitability the following year. At its peak, in 1981, Atlas Consolidated was one of Asia's leading mining companies. Poor prices and poor management, however, left the company in serious financial difficulty by 1984. Financial and managerial problems forced Nonoc Mining and Industrial Corporation to close down operation of its nickel mine and the country's only nickel refinery, in northern Mindanao, in early 1986.

Mindanao's known and suspected mineral resources attracted considerable attention from a variety of interests during the latter years of the Marcos regime. The presence of Muslim separatist and communist New People's Army guerrillas, along with large numbers of small-scale miners, however, made development by mining companies difficult. After beginning exploratory work at one site in 1980, Benguet was forced to suspend further work because of the law and order situation. Apex Mining, in Davao del Norte, found itself reduced to buying waste from small-scale miners who virtually paralyzed its gold mining operations.

Indonesia's mining industry also fell on hard times in the early 1980s, especially nickel mining. In an effort to increase foreign interest in the mining industry, the government introduced a "third generation" of contract in 1976, which increased incentives for foreign companies.[28] However, low metal prices, increasing production costs, and other factors, led many mining companies to pull out of the country or to reduce their operations during the first half of the 1980s. Several smaller nickel operations closed down. After four years of losses, in 1982 Inco sought unsuccessfully to reduce its debts by offering twenty percent of its shares to the Indonesian government and the Bank of Japan. The Indonesian government, itself in financial trouble, did not accept the offer, leading Inco to close its operations temporarily. One exception to this gloomy picture was Freeport, which remained profitable. In fact, it was so profitable that the Indonesian government renegotiated its contract in 1976. The new agreement cancelled the company's tax holiday ahead of schedule and allowed the government to purchase an 8.5 percent share in the mine.

Thailand's tin mining industry, located in the southern part of the country, developed initially largely in conjunction with Malaysia's mining industry.[29] The first tin miners in Thailand were probably Chinese. Technological innovations in the industry were brought by Chinese from Malaysia during the late nineteenth and early twentieth centuries, including the tin dredge in 1906. Significant growth of the industry did not take place until well after the Second World War. By the 1960s, Thailand ranked third in world tin supply (representing fourteen percent of the market) and mining accounted for about nine percent of export earnings. During the 1960s and 1970s, the government actively sought to expand the mining industry. In 1980 minerals were second only to rice as an earner of foreign exchange and it employed around 90,000 people.[30] Unlike other mining industries around the region, however, Thailand's continued to be dominated by relatively small operations. Over eighty percent of the value of mineral output was from small-sized and medium-sized mines,[31] and much of this mining still employed fairly simple technology.

During the late 1970s and early 1980s the Thai government became concerned with internal problems facing the industry, especially encroachment on legal mining properties by poachers and levels of violence and corruption in the industry. Despite such problems, high prices kept the industry buoyant. The crash in the international tin market in 1985, however, seriously damaged the mining industry. Industry earnings dropped to one-half their 1980 level and employment was estimated to have fallen to 40,000.[32] The government responded by reducing royalties and taxes. The industry was also shielded from international problems, to some extent, by growing local demand (which more than quadrupled between 1981 and 1985).[33] Nevertheless, the industry continued to decline throughout the remainder of the 1980s. By the early 1990s only a few tin mining firms were still in operation, and most of these were operating at a loss.

Mining and the Environment

The environmental impact of mining has emerged as a critical issue only in recent years, and even today it is a relatively muted debate for the most part. The national governments of Southeast Asia have viewed mining primarily as a means of increasing foreign exchange earnings. To a lesser extent, mining has been seen as having a role in import substitution, industrialization, and employment generation. Within governments, for the most part, the benefits from mining are judged at the national level, with local considerations being given far less significance, and, all too often, they are closely linked to elite interests—to the interests of the oligarchs, cronies, and so forth that permeate governments in the region. As a result of this centralized and elitist perspective, concern about the environmental impact of mining has not been given high priority by governments in the region.[34] The issue has been raised primarily by a handful of non-governmental organizations, community activists, and academics. Their concern has emerged since the 1970s, largely in tandem with issues raised in other areas, such as forestry, and in relation to discussions of ethnodevelopment and sustainable development.

The Philippines has witnessed sporadic protests over the harmful effects of mining on the environment for over a decade. These initially centered around northern Luzon, where pollutants flowing from mines into rivers threatened ricefields and fishponds.[35] More recently, farmers in the Cordillera have protested plans to develop open-pit mines, such as

Benguet's Tudong Gold Project. While such protests have, in some instances, been associated with calls to reduce pollution and provide greater safeguards against pollution, more often they have taken the form of opposition to any new large-scale mining development. The government's Department of the Environment and Natural Resources is given the dual role of encouraging the development of mining and monitoring its environmental impact, and has encountered considerable difficulty in pursuing the latter role.

In response to growing criticism, those speaking on behalf of the industry have argued that the situation is not as bleak as some environmentalists would have us believe and that the damage caused by mining is less harmful to the environment than more extensive forms of natural resource exploitation, such as forestry. In addition, pressure from the public and the Department of Environment and Natural Resources has resulted in at least a few of the larger companies paying more attention to environmental issues. However, while it might be argued that some of the larger mining enterprises have become environmentally more responsible in recent years, widespread serious problems persist. Moreover, given the move to developing open-pit mines by the industry, such as Benguet's Grand Antamok project and the Southeast Gold Resources project, the environment is likely to feature even more prominently in debate over mining development in the future.

Under Thailand's Improvement and Conservation of National Environmental Quality Act of 1975 (amended in 1978), all mining operations require prior approval of the National Environment Board, based on the acceptance of an environmental impact assessment report prepared by an entity approved by the board. By and large, such provisions have proven ineffective in confronting the environmental damage caused by legal and illegal mining operations in the country. Thus, researchers from the Thailand Development Research Institute commented in the late 1980s that "the mining industry possesses adequate technologies to make ore extraction, recovery and processing a 'clean' business, legally and environmentally. Yet there are widespread cases of 'dirty' operations that have damaged the environment."[36] Public and government concern about the environmental impact of mining has increased over the past few years, and occasional protests against polluting miners have taken place, but little appears to have been done to improve the situation. Even among the environmentally conscious, concern with the mining industry is negligible compared with interest in other forms of environmental degradation and pollution. In the case of tin mining, pollution has been reduced to some extent largely as a result of the decline of the industry.

Indonesia's political system, to a large extent, has shielded large-scale commercial mining operations from criticism. It has also made it difficult to obtain information. Environmental concerns have been included in international criticisms of the Freeport mine in Irian Jaya. International mining companies such as Freeport and Inco, for their part, have taken steps to improve their environmental record, but critics claim that these are still inadequate. Within Indonesia, there remains little direct pressure to make miners accountable.

Small-Scale Mining and the Environment

The discussion of small-scale mining in this section will focus on the Philippines, Indonesia, and Thailand. In the Philippines and Indonesia, small-scale mining is associated primarily with gold, while in Thailand it is linked to tin and gems.

A degree of tension between small-scale miners and large mining enterprises has existed in the Philippines for much of the twentieth century. The 1937 Mining Law of the newly established Philippines Commonwealth banned many forms of small-scale mining which could compete with the larger companies, but considerable scope for small-scale mining remained. The number of small-scale miners in the Philippines increased substantially in the 1980s, primarily in the Cordillera and on Mindanao. The mountainous region of Mindanao to the northeast of Davao City became the scene of a gold rush, beginning in 1982, that attracted tens of thousands of prospectors.[37] The New People's Army and Bangsa Moro Army took over running the mining camp at Boringot in 1984, until they were driven out the following year by the army. Another large mining settlement at Diwalwal received international attention in June 1987, when hundreds of miners were buried as mudslides caused their tunnels to collapse. The number of small-scale gold miners in Mindanao, the Cordillera, and elsewhere in the country has continued to grow since the early 1980s. By 1990 there were estimated to be over 100,000 such miners. During the first half of 1990 they produced an estimated 4,310 kilograms of gold, representing 8.68 percent of the gold produced by volume.[38]

Small-scale alluvial gold mining has occurred throughout Indonesia during this century, but did not emerge as an important political and economic issue until the 1980s. Following devaluation of the rupiah, in 1986 illegal miners descended on northern Sulawesi, central Kalimantan, and Jambi and Bengkulu in Sumatra. Compared to legal gold production in 1987 of 3.3 tons, almost all from the Freeport Mine in Irian Jaya, an estimated 9.1 tons, worth some US$120 million, was exported illegally.[39] In Kalimantan, problems arose with illegal small-scale miners even before the start of commercial activities in the late 1980s. Some 4,500 illegal miners working around the Mount Muro concessions in 1987 were said to be carrying away one ton of gold a month.[40] Estimates of the number of illegal miners in Kalimantan by 1987 were as high as 100,000. Fearing that illegal mining would discourage investors and because of the revenue lost through smuggling, the government passed legislation threatening miners with arrest if they did not cease their activities by 1 July 1988. This excluded local residents, who were allowed to pan for gold. In addition, some local residents had taken out small concessions, which were also threatened by illegal mining activities by outsiders.

Problems with small-scale tin miners in southern Thailand became an important issue in the late 1970s as the government sought to establish greater control over the industry. Most of the tin mining was carried out by relatively small operators, and much of the mining activity was illegal and a good deal of what was produced was smuggled out of the country. As one commentator in the late 1970s noted: "The sheer size of the illegal mining sector and the spin-off wealth it creates, militate against any meaningful attempts to control and regulate it."[41] Another observer, writing in the early 1980s, pointing to the need to enforce mining rights at gun point, the use of small private armies in the industry, and rackets in registering tin dredges, stated that "enforcement of policies have been proven virtually impossible," and noted that attempts to control the industry had only led to more smuggling.[42] In recent years, small-scale tin mining in southern Thailand has declined, being replaced, to some extent, by mining of calcium compounds in Ranong province, and by the growth of the tourist industry around Phuket.

Similar problems of lawlessness characterized gem mining in Chanthaburi and Trat provinces, which border Cambodia. As was noted above, as the gem fields have been depleted, mining has come to be conducted by larger enterprises and small-scale miners increasingly have moved to the gem fields of western Cambodia. Many of the small-scale miners are farmers drawn from the northeast who work in the mining industry in the off-season. The total workforce in gem mining is estimated at around 400,000, many of whom are small-scale miners, and the number of Thai miners crossing into Cambodia is in the range of ten to thirty thousand.[43]

Debate about the activities of small-scale miners generally has taken the form of contrasting them with those of large-scale commercial enterprises. Supporters of small-scale mining pose economic questions in terms of generating revenue for the government versus providing immediate local income, political questions in terms of outside domination versus local control, and environmental questions in terms of the impact of small-scale operations versus the impact of large mines. Such questions relate to support for indigenous rights, opposition to dependency and promotion of local autonomy, and advocacy for appropriate development of the small-is-beautiful variety—and to a faith in the environmental supremacy of such approaches.

Advocacy for small-scale mining is most developed in the Philippines, where it is closely associated with support for indigenous rights and with criticism of the environmental impact of large-scale mining. During the Marcos years, criticism of the mining industry focused on the excesses of Marcos and his cronies and the need to share the benefits of development with indigenous peoples and to promote development that was more in tune with indigenous cultural traditions. The position of the Cordillera People's Alliance (CPA) on economic development was spelled out in 1985. In addition to a demand for "more state allocations and social welfare services for accelerated development" based on the historical marginalization of the indigenous people of the Cordillera from the process of economic development, the CPA argued:

> Any economic development should be within the framework of social justice, the broadest possible participation of the people in the entire development process, adaptation of indigenous forms and systems of collaborative labor and cooperative management, and self-sacrifice. Development strategies shall be geared towards increasing production and improving the simple forces of production in the region.[44]

Joanna Carino, a member of the Cordillera Consultative Committee, contrasting the natural wealth of the Cordillera with the poverty of its indigenous inhabitants, argued: "It is highly doubtful whether the claimed benefit from corporate mining of foreign exchange can ever outweigh the social costs of such a development thrust."[45] She outlined the CPA's "alternative development strategy" in terms of the principles of the right to the ancestral domain, the right to ancestral proprietary rights to the disposition, utilization and management of the natural resources within the ancestral domain, the right to economic prosperity, regional autonomy, and cultural self-determination.[46] Included under proprietary rights was a call for nationalization of all foreign-controlled industries and the payment of compensation to displaced communities.

LUMAD, an organization of mostly non-Moslim indigenous peoples in Mindanao, in a statement issued at its first regional assembly in 1984, argued that under Marcos natural resource development in Mindanao had benefitted only national and foreign business interests at the expense of indigenous peoples. LUMAD therefore demanded

> the dismantling of the monopoly of lands and natural resources from the hands of the foreign capitalists and their local partners,…the return of lands and the indemnification of all Lumads ejected from their lands by mining, logging corporations, plantations and other industries,…the just distribution

of land to the tillers, [and] that tribal lands be protected from the onslaught of dams and other infrastructure projects detrimental to the tribes.[47]

A similar view is held by Mindanao environmentalists associated with the Kinaiyahan Foundation, who argue in favor of sustainable development that recognizes the holistic approach of indigenous peoples: "Tribal Filipinos still recognize the activity of spiritual forces in nature."[48]

As the initial promise of improvement under the Aquino government seemed to disappear, indigenous rights groups renewed their calls for "genuine" development. Accusing the Aquino government of being "in reality, anti-people, anti-democratic and oppressive" one organization noted that "the Indigenous Peoples are not anti-development. They would, in fact, be the first to rally behind any genuine development initiative."[49] In mid-1987, a group of Cordillera organizations criticized a proposal sent by the Aquino government to the European Community for failing to provide an adequate framework for Cordillera development that was responsive to the "wishes of the indigenous communities based on the natural resources of their respective reasons" and "the maintenance of an ecological equilibrium and a coherent and unified model of development for the whole region."[50] Such organizations favor the development of small-scale, community-based mining that utilizes the "simple forces of production" rather than large-scale corporate mining, arguing that such development will be more sustainable and will provide more direct benefits to local people.

The supposed respective benefits of small-scale and large-scale mining became important issues in debate concerning the creation of an autonomous region in the Cordillera in the late 1980s. Decentralization was seen by the Aquino administration as a means of promoting democratization and more equitable economic development. Provision for the granting of greater autonomy to the Cordillera and parts of Mindanao and adjacent islands appeared in Article X of the 1987 Constitution The article allowed for creation of autonomous regions in these areas following approval by a majority of the inhabitants through a referendum. Such a referendum was held in the five provinces of the Cordillera on 29 January 1990. Benguet, Abra, and Kalinga-Apayo voted overwhelmingly against the referendum, in Mountain province the vote was closer but still negative, and only Ifugao province voted in favor.

Reasons cited for the lack of support by those favoring autonomy were the poor understanding of the plan, fear of the unknown, perceived flaws in the legislation, and the influence of local political bosses. There were also widespread rumors that the mining companies had spent a great deal of money campaigning against the referendum since they feared its passage would lead to higher taxation and favored treatment for small-scale indigenous miners. Such factors may well have played a role. Nevertheless, the fact is that those areas with large-scale mining companies did vote overwhelmingly against autonomy. Ifugao, for its part, is a poor province with few mining resources and poorly developed infrastructure. The Cordillera Autonomous Region did not come into being after the referendum and there were widespread doubts about Ifugao's economic viability on its own.

Economically, the problem for the Cordillera is a familiar one. Resource-based mining and timber companies are fundamental sources of government revenue and are important sources of employment. While some argue against their contribution to sustainable development, others fear the negative economic impact should these companies cease operation. Likewise, while some promote future growth based on small-scale, community-based undertakings, others see the only way out of the region's poverty to be by encouraging more investment by large-scale resource-based companies.

While the case for community-based forestry in the Cordillera (and in Mindanao) can be made with relative ease, mining is a very different matter. To begin with, much of the ore that has been mined in this century and that is potentially mineable can only be obtained by intensive, large-scale mining. There are other problems with the small-scale alternative. Small-scale mining certainly provides some opportunity for greater direct community involvement of the type called for by the various groups cited above, but not without also generating serious problems. It can be argued that such mining is more damaging to the environment, less safe for miners, less likely to generate capital that can be used for developmental purposes, and more likely to lead to law and order problems than large-scale mining. Small-scale miners poison themselves and their environment at an alarming rate and the only ones to receive long-lasting economic benefits tend to be those who buy the gold or who provision the prospectors. The Department of Environment and Natural Resources cites the following problems resulting from the activities of illegal small-scale miners: (1) unsafe mining practices and unsystematic mining methods inducing landslides and cave-ins; (2) low value recoveries due to inefficient milling methods; (3) social problems, including poor health and sanitation, and a lack of peace and order; (4) environmental degradation, such as silting of rivers and streams, pollution due to the use of deleterious chemicals, and denudation of forest cover; and (5) unregulated trading of gold resulting in the government's poor recovery of gold produced.[51] The position of the government is to seek to promote both small-scale and large-scale mining, rather than viewing them as alternatives.

Small-scale tin mining has been subjected to some criticism in Thailand. The government's primary concern has been that of illegal mining.

Development plans from 1977-1986 focused on problems in the mining industry, and especially those posed by encroachment on mining properties by poachers—"Of particular concern was the encroachment by villagers onto mining properties held by legitimate owners."[52] Small-scale, illegal tin mining is also generally viewed as environmentally harmful. One critic, Theodore Panayotoy, refers to such mining as damaging "in a most wasteful manner...and a source of pollution that poisons living aquatic resources."[53] For the most part, however, mining has not featured in debates in Thailand concerning the environment or community development.

Conclusion: Development Options

The emergence of the environmental movement in recent years has provided a new dimension to debates over the relative benefits and costs of mining. In the past, issues focused on how to share the revenue generated by mining without fundamentally questioning the costs resulting from environmental degradation. Environmental questions now feature much more prominently in most discussions about mining. Small-scale mining has captured the attention of those most critical of the distribution of wealth from mining and of the environmental degradation caused by large-scale mining as an alternative that is supposedly of greater benefit to local communities and less harmful to the environment. Unfortunately, the latter view has tended to be based more on emotion than careful analysis of what is a very complex situation—romanticizing small-scale mining and uncritically vilifying large-scale mining, especially when conducted by transnational corporations.

Mining companies have received a great deal of criticism from environmentalists, indigenous rights advocates, and other proponents of alternative forms of development. The picture painted by opponents of mining is often of an intrusive and destructive industry with few linkages to a homogeneous, downtrodden local population. The mining companies, for their part, seek to portray themselves as caring and sensitive, providing services and employment, and responsive to rational criticism. While there may be some truth in both perspectives, again, actual situations are far more complicated, especially when trying to assess environmental damage and other developmental questions. While all private sector mining companies are driven by a desire to maximize profitability, they have shown different degrees of willingness to respond to pressure to reduce levels of environmental pollution, to improve the welfare of their workers, and create better relations with neighboring communities. In addition, although the intrusiveness of mining is certainly evident during the early phases of development, once a mine is in operation it increasingly becomes integrated into a complex local setting.[54]

Understanding this local setting is essential to begin to assess developmental options. The equation is not a simple one of miners versus the local inhabitants. The local setting for a mine includes its employees, the residents of surrounding communities, and variety the regional political and administrative entities. The surrounding communities may include people from different indigenous groups or tribal minorities as well as immigrants. The communities themselves can be very heterogeneous, rather than a homogeneous group of simple farmers. They can even include mine workers and their families as well as others who are reliant on the mine in some way. Among those who are farmers, while many may suffer from the pollution caused by mining, some may also benefit from the market and infrastructure provided by the mine. Indigenous or tribal communities may contain such divisions as well as important ones based on kinship or other sociocultural criteria. These divisions, for example, may play a role in debate over whether to oppose mining or to seek compensation, and in determining what constitutes adequate compensation.

The complexity of the local setting is of considerable relevance in assessing the role of small-scale mining in the development process. To begin with, it is important to recognize that there are often important differences from one setting to another. The Filipino view favoring small-scale mining, for example, has developed largely in response to conditions in the Cordillera, where most small-scale miners are from the Cordillera (although not necessarily from the immediate vicinity of the mining). In contrast, in central and western Kalimantan most of the small-scale miners are Javanese or other non-tribal people whose actions, it can be argued, are far more intrusive to the indigenous inhabitants than are those of larger, more concentrated mining operations. Large-scale mines in the Cordillera, such as Benguet's various mines, provide relatively stable and well-paid employment to a large number of tribal people in the Cordillera (a fact usually ignored by those opposed to large-scale mining), while employment for small-scale miners is less stable and riskier. This situation differs from, say, Irian Jaya, where the number of Irianese employed at the Freeport mine, although increasing, remains relatively limited.

Beyond the need to recognize such differences, there are general features of small-scale mining that raise important questions about its benefit to local communities. To begin with, the experience of small-scale mining in most settings is that the wealth produced is of little developmental benefit to the region or even to most of the individuals involved in small-scale mining. In fact, where small-scale mining develops into a "rush" it may have serious detrimental effects in terms of law and order, disruption of other economic pursuits, and to the environment. A related problem is that of regulation. While the power of large mining companies makes it difficult to regulate their actions that are harmful to the environment, they are, nevertheless, susceptible to national and international pressure to behave in a more responsible manner. Such pressure is of little relevance to small-scale miners, who are also very difficult to regulate even by governments.

The conclusion from the above is that we should be very cautious about seeing small-scale mining as an alternative to large-scale mining. It is important to recognize the developmental limitations of small-scale mining for local communities. In addition, while the history of relations between large-scale mining operations and neighboring rural communities has often left a good deal to be desired, such communities may be better served in the long run by seeking to pressure large-scale miners to become better "corporate citizens" than to try to eliminate them in favor of small-scale operations. Such attention to large-scale mining operations does not, however, imply attempting to eliminate small-scale mining. What makes more sense is seeking to establish the best role for both types of mining in a particular setting, recognizing the respective strengths and weaknesses of each one on the basis of careful analysis.

Notes

1.   A descriptive list of the leading mining companies in the Philippines in the 1980s is provided in Michael C. Howard, *The Impact of the International Mining Industry*, (Sydney: University of Sydney, Transnational Corporations Research Project, 1988), p. 68. There is a discussion of ownership of the companies on pages 69-92 of the same work.

2.   Epithermal gold mining in the South Pacific is discussed in Michael C. Howard, *Mining, Politics, and Development in the South Pacific* (Boulder: Westview Press, 1991).

3.   See Richard Gourlay, "Philippines' glister rediscovered," in *Financial Times*, 1 August, 1989, p. 24.

4.   See Jose Galang, "Investment famine blights Filipino mining outlook," in *Financial Times*, 26 March, 1993, p. 36.

5.   On 30 April 1993, Malaysia Mining Corporation, once the country's largest tin miner, announced that it was withdrawing completely from tin production. Its involvement in mining being limited to a 46.95 percent share in Ashton Mining, an Australian gold and diamond mining company. See, "Tin giant quits," in *Far Eastern Economic Review*, 13 May, 1993, p. 75; and Kieran Cooke, "Malaysian group to pull out of tin mining," in *Financial Times*, 5 May, 1993, p. 27.

6.   The contracts signed by foreign companies with the Indonesian government allow them to carry out surveys, exploration, feasibility studies, and building operations, and thirty years of mining. By the tenth year of production, fifty-one

percent of the equity of the operation is to be in Indonesian hands. Most companies began with ten to fifteen percent Indonesian investment.

7. Ownership in Freeport Indonesia in 1989 included a nine percent stake by the Indonesian government and a 3.7 percent share by Norddeutsche Affinerie of Germany. An agreement to renew Freeport's mining contract in early 1990 extended its activities for another thirty years and called for the government's stake in the company to rise to twenty percent.

8. See Kenneth Golding, "Double find leads to about-turn," in *Financial Times*, 5 January, 1993, p. 19; and Kenneth Golding, "Freeport plans Spanish copper expansion," in *Financial Times*, 19 November, 1992, p. 26.

9. See Howard, *Mining, Politics, and Development in the South Pacific*, pp. 68-69.

10. Barbara Crossette, "Indonesia gripped by gold fever," in *International Herald Tribune*, 16 February, 1987, p. 10.

11. Gerard McCloskey, "Indonesia's coal industry ready for the big time," in Financial *Times*, 11 May, 1990, p. 40; see Alan Deans, "Australian mining firms target Asia: Strength in diversity," in *Far Eastern Economic Review*, 12 December, 1991, pp. 60, 62.

12. *Bangkok Post*, 3 January 1992: 17. Padaeng Industry's smelter in Tak produced 72,000 tons of zinc ingots in 1991, all of which was sold within Thailand. In 1990 Thailand exported 5,840 tons of zinc.

13. "Gems and jewellery: Foreign sales regain sparkle," in *Bangkok Post Economic Review Year-End 1991*, pp. 79-80.

14. In late 1992 there were an estimated fifty-seven Thai companies, employing some 100,000 people, with mining operations in Cambodia. One Thai politician estimated that the Khmer Rouge were earning around US$4 million a month from these operations (Rodney Tasker, "Fortunes at risk: Sanctions threaten Thai-Khmer Rouge Trade," in *Far Eastern Economic Review*, 12 November, 1992, p. 13). Also see *Bangkok Post Economic Review Year-End 1992*, p. 79.

15. See *Atlas of Mineral Resources of the ESCAP Region, Volume 7: Lao People's Democratic Republic* (United Nations, Economic and Social Commission for Asia and the Pacific, 1990); "Vietnam's mining revision," in *Far Eastern Economic Review*, 11 March, 1993, p. 59.

16. Sources on the early history of mining and metal-working in Southeast Asia include: Ian Glover, Pornchai Suchitta, and John Villiers, eds., *Early Metallurgy, Trade and Urban Centres in Thailand and Southeast Asia* (Bangkok: White Lotus, 1992); Juan R. Francisco, "The glint of metal: How the Metal Age dawned in Southeast Asia," in A. Roces, ed., *Filipino Heritage*, Volume 1 (Manila: Lahing Pilipino, 1977), pp. 296-302; Tom Harrisson and Stanley O'Connor, *Excavations of the Prehistoric Iron Industry in West Borneo*, Data Paper No. 72. (Ithaca, NY: Southeast Asian Studies Program, Cornell University, 1969); Tom Harrisson and Stanley O'Connor, *Gold and Megalithic Activity in Prehistoric and Recent West Borneo*, Data Paper No. 77 (Ithaca, NY: Southeast Asian Studies Program, Cornell University, 1970); H.E. van Heerkeren, *The Bronze Age of Indonesia* (The Hague: Nijhoff, 1958); Karl Hutterer, "Prehistoric trade and evolution of Philippine societies: A reconsideration," in K. Hutterer, ed., *Economic Exchange and Social Interaction in Southeast Asia* (Ann Arbor, MI: University of Michigan Papers on South and

Southeast Asia, 1977), pp. 177-190; Anthony Reid, *Southeast Asia in the Age of Commerce 1450-1680: Volume One, The Lands Below the Winds* (New Haven, CT: Yale University Press, 1988); William H. Scott, *The Discovery of the Igorots*. Quezon City: New Day Publishers, 1977); Wilhelm G. Solheim II, Philippine prehistory," in G. Casal, *et al., The People and Art of the Philippines* (Los Angeles: Museum of Cultural History, University of California, 1981), pp. 17-84; Lawrence Wilson, *Igorot Gold Mining Methods* (Baguio: Catholic School Press, 1932); Lawrence Wilson, "Primitive mining in the Philippines." *Far Eastern Review*, Vol. 29, No. 12, 1933, pp. 555-558; and Peter Schreurs, *Caraga Antigua: The Hispanization and Christianization of Agusan, Surigao and East Savao* (Cebu City: University of San Carlos, 1989).

17. The few records that exist of relatively intensive mining are by Europeans dating from the seventeenth century to the nineteenth century. It is likely that most of the mines described are not of great antiquity. An early example comes from a Spanish account from the 1620s of an indigenous gold mine at Antamok, in the Cordillera, that was twenty meters deep and employed as many as 800 people. There is a German account from the 1880s of indigenous gold mining in Borneo on the upper Kahajan involving men using ladders and digging shafts, but this technology was probably a recent introduction from Chinese miners. Archaeological evidence from northwestern Borneo indicates the existence of an iron mining and smelting industry, dating possibly from the sixth century, but its more intensive phase appears to date only from around A.D. 1300. There were a few gold and copper mines in Sumatra dating from perhaps the thirteenth century. The technology for these may have been introduced from India.

18. Dutch interest in tin mining in Southeast Asia dates from the seventeenth century, although they did not engage in very large undertakings until the nineteenth century. They began collecting and exporting tin from the Malay peninsula soon after they captured Malacca in 1641. A 1688 treaty with Ayuthia restored commercial relations regarding trade in hides and tin (the French had sought a monopoly in the tin trade in their 1685 treaty). The Dutch were interested in tin deposits in Sumatra and its neighboring islands from the 1680s. They gained access to Bangka's tin deposits from the sultan of Palembang in 1722. See Mary F. Somers Heidhues, *Bangka Tin and Mentok Pepper: Chinese Settlement on an Indonesian Island* (Singapore: Institute of Southeast Asian Studies, 1992); and D.J.M. Tate, *The Making of Modern South-East Asia, Volume One: The European Conquest*, pp. 243, 244, 248, 521 (Kuala Lumpur: Oxford University Press, 1971). By and large, the Dutch bought gold, silver, and diamonds from native producers rather than mine them themselves; see Alex L. ter Braake, *Mining in the Netherlands East Indies*, Bulletin 4, pp. 50, 92 (New York: The Netherlands and Netherlands Indies Council of the Institute of Pacific Relations, 1944).

19. Thomas Stamford Raffles, *The History of Java* (London: Black, Parbury and Allen, 1817).

20. Sources on the development of mining in Southeast Asia in the nineteenth and early twentieth centuries include: Heidhues, *Bangka Tin and Mentok Pepper*; Harrisson and O'Connor, *Gold and Megalithic Activity*; ter Braake, *Mining in the Netherlands East Indies*; J.S. Furnivall, *Netherlands India*. (Cambridge: The University

Press, 1944); Wolfgang Marschall, "Metelurgie und Frühe Besiedlungsgeschichte Indonesiens," in *Ethnologia* (Köln), N.S. Vol. 4, 1968, pp. 29-263; J. Thomas Lindblad, *Between Dayak and Dutch: The Economic History of Southeast Kalimantan 1880-1942* (Dordrecht, Netherlands: Foris Publications, 1988); W.C.B. Koolhoven, "Het primaire diamontvoorkomen in zuid-oost Borneo," in *De Mijningenieur*, Vol. 14, 1933, pp. 138-144; James C. Jackson, *Chinese in the West Borneo Goldfields*, Occasional Paper in Geography 15 (University of Hull, 1970); E.C. de Jesus, *Benguet Consolidated, Inc., 1903-1978: A Brief History* (Manila: Benguet Consolidated, 1978); Winifred Wirkus, History of the Mining Industry in the Philippines: 1898-1941 (Ph.D. thesis, Cornell University, 1974); W.W.L. Blythe, "A historical sketch of Chinese labour in Malaya, in *Journal of the Malayan Branch of the Royal Asiatic Society*, Vol. 20, pt. 1, 1947; Lin Ken Wong, *The Malayan Tin Industry to 1914* (Tucson: University of Arizona Press, 1965); Yip Yat Hong, *The Development of the Tin Mining Industry of Malaya* (Kuala Lumpur: Oxford University Press, 1969); H. Warington Smyth, *Five Years in Siam from 1891 to 1896* (New York: Scribner's, 1898); Jennifer W. Cushmen, *Family and State: The Formation of a Sino-Thai Tin-Mining Dynasty, 1797-1932* (Singapore: Oxford University Press, 1991); also see note 4 for sources on Thailand; Richard W. Hughes, *Ruby & Sapphire* (Bangkok: White Lotus, 1991), pp. 243-244 (on Burma) and pp. 256-259 (on Thailand); T.T. Wynne, "The ruby mines of Burma," in *Transactions of the Institute of Mining and Metallurgy*, Vol. 5, pp. 161-175; H.I. Chhibber, *The Mineral Resources of Burma* (London: Macmillan and Co., 1934).

21. The Spanish and the Americans were primarily concerned with revenue and demarcation issues established mining rules in the Philippines, derived from those in force in their own country. One interesting result of establishing such standards was a ban on women working in the mines. This was of some relevance in the Cordillera, where women traditionally had been involved in mining and were employed in surface-work by some mines. In Indonesia, safety standards were established through ordinances that accompanied the mining act of 1899 (amended in 1910, 1918, and 1930). By and large, Dutch officials appear to have considered conditions adequate. The *Handbook of the Netherlands East Indies* for 1924 (Buitenzorg: Division of Commerce of the Department of Agriculture, Industry and Commerce, 1924, p. 217) noted, in regard to coal mining, that "hygienic conditions are generally favorable among the labourers." In the case of Malaya, labor standards in the mining industry were established following commission reports in 1876, 1890, and 1910. A labor code was put in place in 1912. Thailand established a department of mines in 1891 and enacted a mining act in 1901, but these were little concerned with health and safety standards.

22. Gaining control of the mineral wealth of Southeast Asia had been one of Japan's goals when in invaded the region. However, mining production dropped sharply and, as the war progressed, came to a virtual halt in most areas. In the Cordillera, for example, Benguet ceased operation, but the Lepanto mine was kept open for a good deal of the time since the Japanese were anxious to secure a source of copper for their military requirements; see Howard T. Fry, *A History of the Mountain Province*, p. 191 (Quezon City: New Day Publishers, 1983). The British attempted to destroy mines in Malaya as they retreated, but the Japanese

soon had them operating again. By 1943 production was up to almost one-third of its 1940 level, but sabotage and other difficulties greatly reduced tin production for the remainder of the war: see Nim Chee Siew, *Labour and Tin Mining in Malaya*, Data Paper 7 (Ithaca, NY: Cornell University, Southeast Asia Program, Department of Far Eastern Studies, 1953), p. 9.

23. *The Colonial Office List 1949*, p. 224 (London: His Majesty's Stationary Office, 1949); and Nim Chee Siew, *Labour and Tin Mining in Malaya*.

24. A. Hunter, "Minerals in Indonesia," in *Bulletin of Indonesian Economic Studies*, Vol. 11, 1968, pp. 73-89.

25. On the mining industry in the Philippines in the 1950s through the 1970s, see: Howard, *Mining, Politics, and Development in the South Pacific*, pp. 63-70; John P. McAndrew, *The Impact of Corporate Mining on Local Philippine Communities* (Manila: ARC Publications, 1983); and Raymond F. Mikesell, *The World Copper Industry*, pp. 35-36 (Baltimore: Johns Hopkins University Press, 1979).

26. See Hamish McDonald, *Suharto's Indonesia* (London: Fontana, 1980); Ingrid Palmer, *The Indonesian Economy Since 1965: A Case Study of Political Economy* (London: Frank Cass, 1978); Ross Garnaut and Chris Manning, *Irian Jaya: The Transformation of a Melanesian Economy*, pp. 71-75 (Canberra: Australian National University Press, 1974); and A. Hunter, "Minerals in Indonesia."

27. See Wilson Forbes, *The Conquest of Copper Mountain* (New York: Atheneum Press, 1981).

28. See Hadi Soesastro, and Budi Sudarsono, "Mineral and Energy Development in Indonesia," in B. McKern and P. Koomsup, eds. *The Minerals Industries of ASEAN and Australia: Problems and Prospects* (Sydney: Allen & Unwin, 1989), pp. 198-199.

29. The literature on the history of mining in Thailand is not extensive. See, for example: Cushman, *Family and State;* Chatthip Nartsupha, and Suthy Prasartset, eds., *The Political Economy of Siam 1851-1910* (Bangkok: The Social Science Association of Thailand, 1978), pp. 205-227; Sompop Manarungsan, *Economic Development of Thailand, 1850-1950: Response to the Challenge of the World Economy* (Bangkok: Chulalongkorn University, Institute of Asian Studies, 1989), pp. 144-158; Wolf Donner, *The Five Faces of Thailand: An Economic Geography* (St. Lucia: University of Queensland Press, 1982), pp. 183-194; Suthy Prasartset, "The tin industry as a non-peasant export production of Thailand," in Carl A. Trocki, ed., *The Emerging Modern States: Thailand and Japan* (Bangkok: Chulalongkorn University, Institute of Asian Studies, 1976), pp. 117-130; and Anat Arbhanhirama, *et al., Thailand Natural Resources Profile* (Singapore: Oxford University Press, 1988), pp. 238-282.

30. Arbhabhirama, *et al., Thailand Natural Resources Profile*, p. 246.

31. Arbhabhirama, *et al., Thailand Natural Resources Profile*, p. 249.

32. Arbhabhirama, *et al., Thailand Natural Resources Profile*, p. 246. Mining earned 14,934 million baht in 1980, three-quarters from tin, and earnings dropped to 7,779 million baht in 1985.

33. Arbhabhirama, *et al., Thailand Natural Resources Profile*, p. 13-14. The value of domestic demand rose from 873 million baht in 1981 to 4,187 million baht in 1985, with local production accounting for sixty percent of domestic mineral consumption.

34. I have discussed the issue of centralization and questions related to devolution of power in Southeast Asia in "Ethnicity, development, and the state in Southeast Asia and the Pacific," in A. Pongsapich, M.C. Howard, and J. Amyot, eds., *Regional Development and Change in Southeast Asia in the 1990s* (Bangkok: Chulalongkorn University Social Research Institute, 1992), pp. 70-84.

35. See Howard, *The Impact of the International Mining Industry on Native Peoples*, p. 74.

36. Arbhabhirama, *et al, Thailand Natural Resources Profile*, p. 15.

37. See "Dying for gold," *Asiaweek*, 16 August 1987, pp. 25-30, 35-41.

38. 1990 production figures for small-scale miners.

39. Rowan Callick, "Gold miners lead trading with Indonesia," in *Australian Financial Review*, 2 November 1987, p. 33.

40. Barbara Crossette, "Indonesia gripped by gold fever," in *International Herald Tribune*, 16 February 1987, pp. 7, 10.

41. Peter Fish, in *Far Eastern Economic Review*, 6 October 1978, p. 83.

42. Theodore Panayotoy, "Natural resource management for economic development: Lessons from Southeast Asian experience with special reference to Thailand," in M. bin Yusof and I.H. Omar, eds., *Natural Resource Management in Developing Countries* (Serdang, Selangor: University of Agriculture Malaysia, Faculty of Resource Economics and Agribusiness, 1983), p. 29.

43. *Bangkok Post Economic Review Year-End 1991*, p. 79.

44. "A people's alliance," in *Cordillera Quarterly*, Vol. 2, No. 2, 1985, pp. 12-13.

45. Joanna K. Carino, "Philippines: National minorities and development," in *IWGIA Newsletter*, No. 45, 1986, p. 203.

46. Carino, "Philippines: National minorities and development," pp. 214-219.

47. LUMAD-Mindanao, "In the land of plenty, the Lumads struggle for justice and equality," in *Sandugo*, Annual Issue, 1984, p. 5.

48. "Economic development and the tribal Filipinos," in *Kinaiyahan Bulletin*, Vol. 1, No. 5, 1990, p. 3.

49. KAMP (Katipunan ng mga Katutubong Mamamayan ng Pilipinas), 1990. "Indigenous peoples of the Philippines versus development aggression," in *IWGIA Newsletter*, No. 62, 1990, p. 105.

50. Teresa Aparicio, "Philippines: Organisation and models of indigenous ethnodevelopment," in *IWGIA Newsletter*, Nos. 55/56, 1988, p. 86.

51. Department of Environment and Natural Resources, *1989 Annual Report*, pp. 29.

52. Arbhabhirama, *et al, Thailand Natural Resources Profile*, p. 256. Thailand's mining acts give the government exclusive ownership of minerals and grant it the exclusive ability to grant licenses, while providing little protection for the rights of members of local communities in the face of outside mining interests or giving them adequate recourse to establishing legal mining rights themselves.

53. Panayotoy, "Natural resource management for economic development," p. 28. The Thailand Development Research Institute report (Arbhabhirama, *et al, Thailand Natural Resources Profile*, pp. 275-80) in its outline of key issues facing the mining industry comments on serious environmental problems, but makes no distinction as to the scale of mining operations.

54. Mining can be divided into four primary phases: exploration, construction, mining, and exhaustion. The actual mining phase tends to be the least traumatic for local communities as they come to terms with the industry, while the others are much more so. See Howard, *The Impact of the International Mining Industry on Native Peoples*, p. 208.

5

Mining and the Environment in New Caledonia: The Case of Thio

Donna Winslow

New Caledonia is an overseas territory of France located in the Pacific Ocean, 1,500 kilometers east of Australia and 1,700 kilometers north of New Zealand. It is comprised of one large island—la Grande Terre—which contains the territory's nickel reserves and the majority of the European population centered mostly in the territorial capital of Nouméa.[1] The majority of the native peoples of New Caledonia, the Kanaks,[2] live in the rural areas of La Grande Terre and on several smaller islands—the Loyalty Islands (Ouvéa, Maré, Lifou, and Tiga), the Bélep Archipelago, the Isle of Pines, and Huon Island. The territory is divided into three provinces, the South and the North (on La Grande Terre) and the Loyalty Islands.[3]

When New Caledonia was discovered by Captain Cook in 1774, he found the islands occupied by a Melanesian people scattered along river valleys and the coast in small hamlets. It was this social space of family residences, agricultural lands, water channels, hunting and gathering territories, which formed the basis for ritual, economic, political, and social action in traditional times. After New Caledonia was annexed by France in 1853, the development of the colony became tied to settler colonialism, mineral exploitation, ranching, and the establishment of a penal colony, all necessitating the expropriation of large tracts of Kanak land and the subjugation of the Kanak people. The dispossession of the Kanaks of their rights over their lands in New Caledonia has followed a classical pattern of colonial expansion and the development of unequal socioeconomic relations which permitted the exploitation of labor, land, and natural resources, all justified by models for economic development imposed by the French state.

New Caledonia was transformed by foreign capital and the growth of an imported labor force recruited to suit the needs of the growing colony. Resources were channelled into the production of primary commodities, such as nickel, for export in exchange for goods and services from the metropole. The nature of the commercial and financial relations between the territory and France served to promote the growth of the capitalist sector at the expense of the subsistence sector, in which the majority of indigenous people lived, and increased the dependency of the territory.[4] Agriculture and cattle ranching play a minor role in the economy of the territory, although an estimated thirty-four percent of the population is still involved in these activities. The domestic production of maize, wheat, rice sorghum, yams, fruits, and vegetables is insufficient to meet the needs of the population and each year the territory imports cereals and other food products. Ranching produces only two percent of the territory's GDP yet thirteen percent of La Grande Terre is devoted to livestock.

The basis of New Caledonia's economy is the nickel industry, which accounts for approximately eighty percent of the territory's exports. New Caledonia is a rich source of nickel. It contains twenty-eight percent of the world's oxidized nickel deposits and it is the third largest producer of nickel in the world. The territory's economy is heavily dependent on nickel production and the industry's boom and bust cycle has shaped New Caledonia's development since the mineral was discovered in 1873.[5] Apart from the civil service, the nickel industry is one of the territory's largest employers, hiring approximately 3,000 skilled and semi-skilled laborers. When the nickel industry went into a recession in the mid-1970s with the fall in world nickel prices, the effects in New Caledonia were devastating: unemployment, social unrest, and economic stagnation.

Nickel is certainly "an affair of state" in New Caledonia. During the 1960s, French authorities made constant reference to just how important they considered New Caledonia's nickel industry to be for France. According to Howard, "They made it clear that it was a primary reason for France retaining control of the territory and emphasized that nickel mining in New Caledonia should, first and foremost, serve the interests of France over those associated with other countries."[6] Thus, in a 1965 speech, President de Gaulle of France stated that the French state should retain control over the nickel industry "in order to preserve French independence in the world economic system."[7]

Such statements were accompanied by a more direct involvement in the mining industry by the French government. Government regulation of the industry was formalized in January 1969, with enactment of the *lois Billotte,* which gave the French ministry of industry power to authorize mining and to set mineral export quotas for New Caledonia. By 1974, the French government had acquired a fifty percent interest in the Société le Nickel

(SLN), the territory's largest nickel company, through the Société Nationale Elf-Aquitaine (SNEA). The other half remained with IMETAL, which was largely in the hands of the Banque Rothchild and the Banque de l'Indochine et de Suez (which held a monopoly on credit in colony). When these banks were nationalized, the French government became the owner of the SLN and thus a major player in the territory's nickel industry. Controlling New Caledonia assures France a regular supply of nickel, which is now considered a 'strategic mineral' since it is an essential element in the production of weapons, electronics, aircraft, and nuclear energy, let alone car bumpers and beer cans. By its continued presence in New Caledonia, France reduces its dependence on foreign sources and has control of a large proportion of world nickel reserves.

In April 1990, Louis Le Pensec, the minister for overseas departments and territories, pointed out the various ways in which the overseas possessions were of value to France. These included the geostrategic value of the nuclear testing site in French Polynesia and the missile testing site in French Guiana. In addition, the overseas possessions were of present or potential economic value. New Caledonian nickel was "in sufficient quantities to constitute a certain economic stake for metropolitan France."[8] New Caledonia's dependency on the production of nickel as its single most important export product makes the territory particularly vulnerable to fluctuations in the world market and the international interests involved in nickel production are not linked into the New Caledonian economy.[9]

Environmental Impacts of Mining

Opencast nickel mining began in the territory at the end of the last century. The nature of nickel extraction represents a serious threat to the physical environment since it destroys the natural vegetation and strips away the surface layer of soil, leading to severe erosion on the steep slopes of the mining areas. The nickel ore is found on the ultramafic rocks in a zone of concentration lying under a layer of weathered material that can be up to thirty meters thick. Large-scale extraction on the mountain ridges and plateaus is done by scraping off this surface layer. Extraction proceeds in horizontal cuts forming giant benches five to eight meters wide. In 1943 and 1954-1955, exploitation was confined to the summit ridges where maximum ore concentrations were more easily reached. Subsequently, the whole central portion of the mountain masses was cut off from four hundred meters above sea level upwards. Nickel ore is extracted at increasing depths because of depletion of the rich veins, which means that ever greater volumes of tailings must be disposed of. In less than one hundred years, 110 million tons of ore have been extracted, resulting in the mobilization of a mass of waste at least five times greater by weight, its volume being between 220 and 280 million cubic meters at the very least.[10]

This mining waste has had serious impacts on New Caledonia's natural environment. Nickel deposits are located in the upper portions of mountain masses, these mountains are extremely rugged with few natural sites for *in situ* disposal of tailings, very steep sides right to the foot of the mountains, tracks and roads are difficult to establish, and loose materials cannot stay in place. Roadworks, access tracks hastily bulldozed out to mining or prospecting sites, have also given rise to intense erosion. The scraping away of the surface earth which is necessary to reach the ore, directly destroys the vegetation. The resulting loose earth is all the more vulnerable to erosion by water. This is intensified by strong seasonal rains and cyclones. And when there is a spell of dry weather, heavy dust is produced by mining operations, especially by the haulage trucks. Up in the mine, this dust is detrimental to workers' health and safety. Down in the plains, it affects dwellings and farms. Pasture lands become less fertile 100 to 200 meters on either side of the road and cattle shun this area.[11]

Huge amounts of loose materials have been washed into the valleys by water, clogging the minor bed of streams and thus causing flooding in the major river beds and covering the fertile agricultural lands of the valleys. In the river valleys, the loose material forms thick deposits, burying the vegetation on the banks and often killing it as well as the animal life in the water. It raises the bottom of the main water channel and thus causes repeated flooding in the flood plain and in the downstream parts of the water courses. Systematic dumping of waste into the natural drainage system spreads pollution down to the marine base line. The accumulation of fine earth particles washed into the sea, especially during heavy rains, damages the coastal flora and fauna. However, the long-term damaging effects of such pollution, very obvious in the case of corals, have not yet been well identified. According to Dupon, "the existence, around the main island, of a barrier reef enclosing one of the largest lagoons in the world has heightened the risk of environmental damage by allowing continental deposits to accumulate in generally calm and shallow waters."[12]

In the 1980s geographers Bird, Dubois, and Iltis noted, that compared to other mining areas elsewhere in the world, "the impact of opencast hilltop mining in New Caledonia has been exceptionally severe and extensive."[13] According to Dupon, New Caledonia offers a "spectacular example of environmental damage resulting from intense and uncontrolled mineral exploitation."[14] Certainly the territory offers a striking example of the damage that can be done to a Pacific island environment by poorly controlled mining activity. Moreover, the damage is not only physical but social. Nickel mining has affected all areas of life in the territory and in the following section I will look at the specific impacts in the commune of Thio.

The Case of Thio

New Caledonia contains thirty-two communes or districts. Many of them host agricultural, urban and industrial activities, however, in each commune one of these activities dominates the lives of its inhabitants. According to Doumenge,[15] five communes are dominated by mining, of which Thio is one. Located on the east coast of la Grande Terre, northeast of the territorial capital of Nouméa, Thio is 997 square kilometers with slightly more than 3,000 inhabitants.

Mined since 1876, twenty-nine percent of the territory's nickel production has come from the deposits of the Plateau de Thio and the adjacent Thio, Mission, Ningua, and Ouenghi mines. The Thio Plateau contains the largest nickel deposit in the world.[16] It has been mined continuously since 1901 and in nine decades has yielded approximately twenty million tons of ore, producing about 450,000 tons of nickel and cobalt.[17] Thio is an example of how macro-economic forces related to international mining interests can have a profound effect on the ecological and social environment of a small community.

For more than a century opencast mining in Thio has produced twenty million tons of nickel, which implies the production of over one hundred million tons of waste material[18] disposed of by mining companies and miners. The Thio mountain sides are seared by opencast terraces, little forest is left, and navigation and fishing in the Thio river is no longer possible.[19] According to Dupon, the Thio center provides, especially by its plateau mine, "the most striking example of environmental damage caused by mining activity in New Caledonia."[20]

The postcontact history of Thio is meshed with that of mining. European colonization of the region and land expropriations can be linked to the needs of the mining industry. Native reserves were introduced to the region in March 1880. By July 1887, the SLN had received authorization to build a tramway line across the Kanak reserve at Ouroué. According to the Marist missionary at the time, "the SLN, rather than compensating the Kanaks as arranged, declared the tramway line 'of public interest' and expropriated the lands it needed."[21]

Two years later, the SLN had taken over two hectares of reserve land in Ouroué for a smelter . On paper,[22] only two of 116 hectares were ceded, but in reality, the presence of the tramway combined with smelting activities forced the Kanaks to abandon the reserve altogether. Ten years later only two Kanaks were still living on their land.[23] This is ironic since the smelter only operated a short time. Inspired by methods used in copper smelting, the SLN built a fusion smelter in 1885 in Le Havre and a few years later another one in Thio. However, the discovery and exploitation of rich nickel deposits in Canada in 1889 led to overproduction and a crisis in the industry. The SLN had to reduce its production in 1891, closing the smelter in Thio after only two years of operation. While smelting continued in the SLN's European operations, New Caledonia was reduced to being a producer of raw materials.[24]

In 1892, the Kanaks of Ile Sainte-Marie in Thio had to evacuate their reserve because of the needs of the SLN and the threats of the head of Indian affairs, Mr. Gallet. Counselled by the local priest, they accepted a small indemnity for the loss of their huts and coconut palms.[25]

> For some time now, the natives have been very upset with the Company "Le Nickel" because it is infringing on their lands here and there, and even though the Company offers an indemnity for the loss of their crops, it is not enough to compensate for the loss of their lands. I was meaning to write to you about this your Emminence, when one morning Mr. Gallet came to inform me of what they were intending to do with Sainte-Marie. I told him that I was totally opposed since the inhabitants of the island have always been good allies of the Europeans even though their distance from the mission has meant that they were not always behaving as they should....
>
> That night Mr. Gallet and Mr. Grand, Mr. Vernay and some engineers went to Ste-Marie...The natives refused to give up their lands and Mr. Gallet had to flee the reserve under their threats. The next evening the natives came to tell me what had happened and I counseled them to submit since, whether they wanted to or not, they would have to leave Sainte-Marie, and it was best to get as much compensation as possible.[26]

Colonization upset what was already a precarious balance between the inhabitants of the region and their natural environment. Land expropriation led to a reduction in traditional farming. According to Dauphiné, reserves in Thio bordering a large SLN mine were defined according to the principle of three hectares per inhabitant even though traditional farming methods are estimated to need twenty-seven hectares per person.[27] In 1900 one reserve in Thio occupied 759 hectares. With a population of 250 people this is 3.03 hectares per inhabitant. Dauphiné goes on to note that there was no real reason to put the natives on the reserves other than to meet the growing needs of the mines for land.[28] Regardless of whether the Kanaks were allies or enemies they received the same treatment concerning land expropriation since only the interests of the mine or European settlers were taken into consideration by the colonial administration. In a letter to the archbishop, one missionary complained, "Mr Frey of the SLN recognizes that the contract negotiated with the natives two or three years ago is only for use of the lands...These lands were given to the natives who aided France in the repression of native rebellion of 1878 as their inalienable right by the Administration....In 1901 Governor Feuillet secretly gave these lands away. It is unjust and illegal...The SLN is now using the land as a maritime zone."[29] Thus, the first impact of the mines on the Kanaks of Thio was the loss of their lands.

Land expropriation also affected traditional farming practices. Yam and irrigated taro production (both "noble" items exchanged in rites and ceremonies) declined dramatically due to the reduction of fallow periods caused by increased population density on some reserves. The adoption of new crops, such as manioc and New Hebridean taro, permitted the Kanaks to survive in these new conditions, but upset the social relations of production associated with traditional farming methods. The introduction of cash cropping—coffee production in almost all the reserves and copra production in the coastal areas—and wage labor in the mines only accelerated these changes.[30] A native head tax was introduced in 1895, applied to males over sixteen years old. The need for cash in order to pay the tax forced the natives into wage labor, although some had been working for the mine since the beginning. The letters from Father Ameline in 1890 note that: "the directors of the SLN think that the natives of Thio should work without rest and in large numbers at the mines...The natives have six month contracts at the SLN but when the planting season arrives they have to leave their work and do so without pay since the SLN says 'finish your six months and after you will be paid,' and thus they lose their money."[31]

By the early 1900s Thio had become SLN's largest mining center[32] and headquarters since its proximity to the coast permitted easy transport. Offices, a school, a hospital, tramways, a monorail, docks, warehouses, and plants were all installed in the region.[33] In addition to its own mining operations, the SLN began negotiating contracts with a certain number of individual mine owners in the region.[34] Because of this and its age, the Thio plateau has a very complex history of mining concessions compared to other areas of New Caledonia.[35]

At this time the industry began experiencing labor shortages due to the cessation of penal transport to the colony. Negotiations with the Japanese government led to an agreement which supplied Japanese laborers to the mines of New Caledonia for twenty years.[36] In 1888-90, five hundred Japanese workers were imported for the mines of Thio. The use of imported contract labor added to the demographic pressures in the region and led to new difficulties with the native population although we only hear echoes of it in the archival material.[37]

The second important impact (on Kanak agriculture) was pollution from the mines. The mines themselves are in mountainous areas of no real agricultural value to the Kanaks and if there had been no side effects from nickel production the two activities might have coexisted, however, mining activities in the mountains severely impacted agricultural activities in the valleys because of flooding due to erosion. The mining materials have choked up the rivers with many alluvial islets and have deposited sediment along the banks so the development of agriculture has been severely limited and the production of traditional tubers, which rot easily when humid, has been disrupted.

Substantial masses of sediment have slumped down the hillside and have been washed into the Thio river. The river which was navigable during precontact times for a length of twenty kilometers is no longer so due to four to five meters of waste deposited along the river bed.[38] Thus, the surface of cultivable land is limited. According to Tissier, the risk of flooding caused by nickel pollution constitutes the most limiting factor for the development of Kanak agriculture in the region.[39] In addition, local climatic factors aggravate the topographical problems. It is difficult to control erosion on the steep slopes in a zone where cyclones can provoke heavy rains. For example, five hundred millimeters of rain fell within twenty-four hours at the Camp des Sapins in Thio during cyclone Alison and rainfall of fifty millimeters within a quarter of an hour is not unusual in the region.[40]

After the original growth of 1875-1900, several surges in production can be noticed in 1901 and 1907 which correspond to the opening of several new mining centers in the south of the territory and there was prosperity in mining until the end of the First World War.[41] The SLN decided to begin smelting again and in 1908 decided to open a fusion plant in Thio near its most important mines. The company opened several new mines deep in the valley and linked them to the coast by a railway in 1910. The amount of ore transported was so great that the company began constructing a twenty kilometer monorail in 1912 to take the ore from the interior to the coast where natives worked loading and unloading at the docks.[42] Within a year, the SLN began producing nickel matte in the new Thio smelter.[43] This soon led to pollution problems and the local Catholic missionaries began to complain. In order to calm them, the SLN began giving money to the mission:

> Because the winds carry the smoke and noxious gases from the smelter, the SLN which wants to maintain good relations with the Mission send Father Dumussy 100 Francs each month. However, this large amount of money does not come close to compensating for the inconvenience caused by the

smelter, which cramps the village and the constant aggravations which this vast exploitation causes the Father.[44]

According to E. Bird, J-P. Dubois, and J. Iltis, nickel metallurgy has accompanied mining since its inception, the remoteness of European markets justifying the smelting of the ore within New Caledonia.[45] The first nickel smelter began operating at Pointe Chaleix, Nouméa in 1879 and two other processing plants were subsequently established: one by the Société des Hauts-Fourneauz de Nouméa at Doniambo, Nouméa, in 1910, the other by SLN at Thio in 1913. A visiting missionary to the area describes the settlement:

> The Thio mission is only two hectares. The catholic native population numbers 367. There are many Europeans and Asiatics...The Société le Nickel, against all laws has obtained from Governor Feuillet almost all the farming land belonging to the natives of Thio, even their cemetery, and has pushed them to the foot of the mountain, onto rocky land, about a hundred meters from the loading dock and the smelter. Thio station thus forms a small green space in the middle of these buildings and yellow earth. It's an oasis of solitude in the middle of all the noise and stands in sharp contrast to all these marks of civilization.[46]

As earlier, the smelter caused pollution. In the letters of Father Dumessy there are constant references to the problems and complaints of the Kanaks regarding the noxious gases and dust from the processing plant.[47] However, Father Dumussy's protests were of little avail and in the early 1930s the roof of the church dissolved due to sulfuric emissions from the plant.[48]

The processing plant in Thio closed in 1931 when the nickel smelting plant at Doniambo passed into the control of SLN. There was a rationalization of the nickel industry in New Caledonia at the end of the First World War boom caused by the end of the arms trade, devaluation of the franc, and competition from other international companies such as Inco. The SLN responded with belt-tightening: it closed its refining factories in England and reduced production in France, concentrated its staff (including administration) in Thio, and abandoned a certain number of mines. It also bought out a concession from a hydroelectric plant at Yaté and in 1926 the SLN completed construction of the plant and opened up an electro-metallurgical factory. New Caledonia was by now the largest supplier of nickel in the world and the SLN was a major enterprise in the colony.[49]

During the interwar years, nickel production in New Caledonia (mainly coming from mines at Thio, Bourail, and Koné) averaged around 150,000 tons of ore per year. In 1920, nickel and chromite accounted for forty-seven percent of New Caledonia's exports by value.[50] Labor became more and more of a problem and Javanese and Vietnamese laborers began to replace the Japanese in the mines.[51] From 1921-23, the SLN began to concentrate its activities in the Thio area. This period was the most depressed for nickel production in the region and the SLN registered major losses until 1924-25, when the market began to pick up again.[52]

By the 1930s, the SLN was reinvesting in Thio. The monorail and dock were repaired. Two new mines were opened and the company began signing contracts with local miners which caused a boom in prospecting in the region.[53] In 1931, Société Caledonia merged with SLN, which thus acquired control of mining concessions covering a total of 150,000 hectares.[54] The production boom of 1939-41 allowed the SLN to stock nickel just prior to the Second World War. For the first time more than 10,000 tons were produced. During the war the island was cut off from Europe and occupied by American troops. Exports to Japan were stopped. This could have led to an economic depression in the colony, however, the Americans supported the production of nickel which was seen as essential to the war effort. The Americans even went so far as to stop a strike against poor labor conditions organized by Vietnamese mine workers.

There was a depression following the war, but in the 1950s reconstruction led to new demands and production reached a new peak in 1953 at 17,037 tons. Because of the labor difficulties during the war, the SLN began to mechanize its production and the face of mining began to change. Manual labor (pick and shovel) which, nevertheless, permitted Thio to produce between 1,000 and 1,500 tons of nickel annually, was replaced by bulldozers. This led to an increase in productivity and the territory produced almost two million tons between 1950 and 1970 and 6.6 million tons between 1970 and 1976. By this time the mine fields were being exploited by giant machines (large bulldozers, seventy ton hydraulic shovels, forty ton trucks, etc.).[55]

The mechanization of mining greatly increased the rate of waste production and downslope movement of discharged debris into the river valleys below. As a result, the banks and bedload of the lower Thio consist entirely of sand and gravel washed down from the Thio plateau. An alluvial island 400 meters long has been completely blanketed by this debris, its scrubby vegetation destroyed. The Thio delta has undergone rapid changes as a result of increase in the solid load carried by the river. Continental deposits progress to the north of the delta along the beaches in the direction of the coastal currents. The fine clay particles settle in the lower part of the river and on the sea bottom. Along the coast, however, the effects of this on the aquatic flora and fauna are yet ill-defined.[56]

The whole upper portion of an ultramafite mountain mass, in contact with the volcanic and sedimentary materials of the central range, where the Plateau-Carrieres (Quarries) mining complex occupies about 1,600 hectares, has been scraped away for the extraction of the nickel ore. The local vegetation has great difficulty in recolonizing the devastated areas. The slopes of the Thio plateau have several areas where mining waste has spilled down and caused large landslides.[57] According to Tissier, the Koua valley has been overwhelmed by thousands of tons of debris from the Camp des Sapins.[58]

The DoThio drains a small basin northwest of the Thio that has a headwater region of volcanic and sedimentary rocks and several ultrabasic massifs near the coast where hilltop mining is active:

> North of the mining area, the valley of the DoThio river also receives, from several ravines running down from the plateau on its right bank, considerable quantities of waste. Other mines opened on the ridges of the mountain mass lying north of the DoThio valley also feed waste materials into this river from the left bank. As on the Thio, several sand and clay banks that clog its mean water channel, the extension of the little delta of its mouth and the pollution of the coastal waters by suspended deposits may be observed. On the right bank, the accumulated materials that have been dumped below the Carrières mines and those that have filled in the Ouanamourou valley, below the Plateau mines are striking.[59]

An attempt was made to control the discharge of coarse waste material down the valley of the Wellington, tributary to the DoThio, by bulldozing boulder levees and a diversion dam, but floodwaters continue to deliver some of the finer material to the DoThio. The mouth of the river threads through a beach ridge plain to a small sandy delta, off which there are shoals deposited from floodwaters in a sea that is stained red by suspended clay in the nearshore zone.[60] Not only is damage caused by the mines but also by prospecting activities and the construction of roadworks to transport the ore to sea. Unfortunately, the location of the ore in the soils of La Grande Terre prevents a different mining technique from being used.

The social environment of Thio also supports the mining industry. The repatriation of Vietnamese workers in the 1960s led to a labor shortage. Polynesian immigrants could not fill the gap and the local Kanaks were called in to meet the demand in the industry as unskilled laborers.[61] The proximity of Thio village, the mining center and the administrative center where the majority of the jobs are concentrated, combined with the small amount of available land (i.e., land that does not flood) for agriculture has destructured subsistence agriculture in the region.

An examination of the demography of Thio shows that the town itself is primarily European, surrounded by a zone containing the Thio mission and the SLN headquarters of Petroglyphe, that contained approximately 1,320 non-Kanaks in 1983. This is before violence in the region in 1984 which led to the departure of some of the European residents of the town. The non-Kanak population works in the town in commerce, professional and service activities, administration, or for the SLN in its administrative headquarters, as skilled labor in the Plateau and Camp des Sapins mines, and at the docks. The surrounding Kanak tribes of Ouroue, St-Philippo I and II, St-Paul, St-Michel, and St-Pierre are interconnected by roads.[62] The majority of the Kanak population lives on reserves and, according to Tissier, in 1985 there were 555 Kanaks living in these reserves (eighty eight men, one hundred and five women, eighty-three youths more than eighteen years old living at home, and 279 children less than eighteen years old).[63] The few European farmers work part-time in agriculture and part-time in salaried employment. As agriculturalists they are often in conflict with local Kanak over land.[64]

The productivity of labor in agricultural activities, that is the amount of wealth produced in one hour of labor, does not allow the agricultural sector to compete with mining. At the SLN salaries are particularly high and disproportionate compared to agriculture. Under these conditions young men have little other desire than to obtain a job in the mine.[65] Mining thus sets salary standards and establishes expectations. It is surprising to note that even commercial fishing, which has the possibility to be financially competitive with mining, does not attract the young men of Thio.[66]

There have been several projects aimed at encouraging young men to return to agricultural activities, but the young men abandoned the projects the minute a job opened in the mine. A pig farm project in the Koua valley failed due to "a lack of motivation" which can easily be explained by the salary difference between a job at the SLN and revenues from agriculture and animal husbandry.[67] According to Tissier, the agricultural economy of Thio has been destabilized by mining since 1876.[68] Not only does pollution cause flooding, therefore reducing the areas of arable land, the salaries inhibit the development of commercial agriculture such as coffee plantations, which have been successful in other areas of La Grande Terre. The economic interest of each producer in agriculture, animal breeding, or fishing can be measured, generally, in terms of the estimated profit compared to other outside activities. This estimated profit, compared in Thio to mining, can be high for certain non-mobile members of the Kanak population such as elders and women, but is totally uninteresting for young men who are free to migrate to the mine fields.

The traditional Kanak social structure of redistribution and subsistence agriculture assured by elders and women, permits these young men to wait unemployed on the reserve for a job to open in the mine. "We practice a form of social welfare with our young people. But what to do? We can't throw them off the reserve and let them join the ranks of unemployed bums in Nouméa."[69] Thus, the remnants of the traditional system support the boom and bust cycle of nickel mining by maintaining a reserve pool of labor.

Kanak SLN workers are on the average young men. 55.6 percent are less than thirty five years old. 62.7 percent of office workers are less that thirty years old, fifty percent of skilled laborers are less that thirty-five years old. In the mine fields, seventy percent are between twenty-five and forty-five, while workers in the processing plant are even younger. Mine work is mainly carried out by unskilled thirty-five to forty-five year old men from La Grande Terre.[70]

The Kanak labor force in the mines is also predominantly male, thus reinforcing a sexual division of labor where women stay on the reserves with the children or work as domestics in the homes of the Europeans while the men hold a salaried employment. However, work in the mines does assure a higher standard of living for families than for those involved solely in agriculture.[71] As early as 1969, the national census bureau had noticed that the revenue of a Kanak family living in a mining zone was three times higher than that of one living in an agricultural zone.[72] In addition to the higher income, families of wage earners in mining benefit from family allowances. Through this system, a father of a large family can double his income. There is an extra financial reward if only one person of the couple works. This too encourages women to stay in the domestic sphere in order to qualify for the additional revenue.[73] However, it is often the case that the only real source of family income comes from the allowance checks while the men spend their salaries on conspicuous consumption:

"Working outside of the reserve is undoubtedly more lucrative for a man than vegetating in agriculture. This realization quickly leads to the temporary or permanent departure of a man from the reserve. However, it is important to note that his future is often influenced by the satisfaction of modern needs derived from daily contact with the European world. Certainly, men work in order to build a solid home out of modern construction materials, but they also spend their money on cars, motorcycles, transistor radios, clothes and gadgets, or on substantial gifts during customary exchanges which make them the envy of the reserve community. However, too often wages earned outside the community are spent on abusive consumption of alcohol."[74]

Alcohol was introduced to the region by Europeans as a trade item. French permissive cultural attitudes towards its consumption have aggravated the problems in the reserves and alcohol abuse by men has led to domestic violence. French wines are easily available in corner stores and many Kanaks die from alcohol-related traffic accidents.

The problems of alcohol abuse are not recent in Thio. They date back at least to the earliest contact with wage labor. In 1890 the Thio Catholic mission wanted to open up its own native store so that the Kanaks "would not do business with the European houses of drink."[75] The mission records contain numerous references to the evil effects of alcohol consumption in the reserves and the presence of night clubs nearby. An extract from a mission diary in 1949 describes the same phenomena noted by Doumenge above twenty-five years earlier:[76] "The SLN has installed a mineral loader which employs most of our young men. If their pay was well spent, it would be wonderful for them and their families. Unfortunately, many of them spend the money uselessly on alcohol and gambling while their children are dressed in rags. Another consequence is that their night work affects the number of communions we can hold."[77] This diary entry also brings up the question of shift work. Although there is no information available on the subject, certainly shift work has consequences for social organization and agricultural production and the role of women on the reserves.

The employment of Kanaks in the mines was in response to expansions and contractions in the industry due to international conditions. From 1965 to 1971 there was unprecedented prosperity until the market took a turn for the worst in 1972 because Japan (who previously absorbed forty percent of new Caledonia's production) began to diversify its sources of nickel to cheaper sources in the Philippines and Indonesia and because of a world economic crisis which meant slow growth in western countries. This influenced the demand for nickel and consequently for labor in the mines. Employment levels in the mines fell from 1,567 to 445.[78] The downturn in the industry led to an economic crisis in the territory. Kanaks were particularly affected by the lay-offs and political tensions began to mount as the Kanaks began asking for a bigger share of the economic pie.

The violence in the 1980s was part of a general Kanak uprising against colonialism in the territory. The Thio Kanaks participated in a territory-wide election boycott which was very successful in the region—only ten Kanaks and less than twenty-five percent of the 1,700 inscribed on the electoral lists (that is 541 Europeans) voted. The Kanak liberation front (FLNKS—Front de libération nationale kanake socialiste—a coalition of the Kanak political parties that favors independence from France) had also been highly critical of SLN on a number of issues during the disturbances. One issue concerned the lack of employment for Kanaks and the treatment of those who were employed. Out of a total of 2,300 SLN employees, only three hundred were Melanesian.[79] In November 1984, armed militants from Canala and Thio invaded the east coast mining village of Thio, blocking the access roads, setting up maritime surveillance and confining the European, Wallisian, French Polynesian, and other non-Kanak families to their homes in a state of (mostly) non-violent siege for three weeks.[80]

During the three weeks of occupation Thio was a symbol of Kanak defiance of the colonial order. It was not only young male Kanak militants who participated in the occupation but the majority of the communities in the surrounding areas. Traditional elders, women, and children could be found providing logistical support for those manning the barricades.[81] The take over of Thio was important symbolically to the Kanak independence movement since the presence of so many non-Kanaks in the area was due to the presence of the mines. Moreover, the majority of Europeans in Thio supported right-wing organizations and the mayor was linked with the extremist National Front party. It was the only commune on the east coast administered by a European, and one who was a confirmed anti-indépendantiste.[82] The Kanaks openly defied colonial order, neutralizing a mobile gendarme unit sent in by helicopter to dismantle the barricades and escorting them back to the local barracks, disarming non-Kanak residents, and seizing weapons in car and house searches.

The Kanak militants managed to close down the SLN and paralyze all activity until an agreement with the French government was reached and barricades were raised in December. According to Gabriel and Kermel, this marked a turning point in FLNKS strategy, that is, the change from armed confrontation to economic destabilization.[83] The boycott of the SLN was maintained. Kanak workers continued to strike and the FLNKS forbid mining operations in the region. Some FLNKS members wanted to extend the strike to a general one in the territory but the independence front lacked the means to enforce its will and the front eventually opted to organize a parallel Kanaky with its own economic networks. In the end, this new attitude helped the territory "normalize" its activities. During his visit in January 1985, President Mitterrand promised that the Thio mine would be reopened soon, but moves to reopen the mine prompted rightwing extremists (anxious to prompt further French military intervention) to set fires and use explosives against the facility, causing A$3 million worth of damage and delaying its opening. The value of lost production between November 1984 and February 1985 was estimated to be around A $15 million.[84] "Thio is the symbol. Nothing must work here.. the situation is abnormal and it must remain so until we obtain satisfaction."[85]

However, normalization progressed and Thio was seen as an experiment in dialogue. The local FLNKS committee wanted one-third of all management positions at the SLN to be occupied by Kanaks. Negotiations followed and the boycott ended in April 1985. The September 1985 municipal elections were won by the FLNKS, which was now standing as a political party.[86] By the end of 1985, the non-Kanak population of Thio had declined and the SLN took steps to promote better employment opportunities for Kanaks.[87] The number of Melanesians employed at the Thio mine itself rose to ninety-five out of a total workforce of 243 by the latter part of 1986, and one Melanesian was promoted to foreman and another sent to France by the company to train for a managerial position.[88]

The tensions in the territory continued for the next few years and, by 1988, violence had increased dramatically causing loss of life on all sides. In an effort to avoid civil war, France's prime minister, Michel Rocard, brought together members of the FLNKS and the RPCR[89] to decide the future of the territory. The results of these negotiations are known as the Matignon accords and they herald a ten year "peace period" during which the French government will attempt to redress the socioeconomic inequalities in the territory, particularly by promoting development and training programs in Kanak communities. In 1998, at the end of this ten year period, New Caledonians will be asked to choose between independence and staying within the French Republic.

In terms of mining, some of the major changes are that the SLN is investing heavily in an ambitious program (1.2 billion french francs) to increase the capacity of the Doniambo smelter in Nouméa from 46,000 to 50,000 tons (of nickel content). In order to meet the smelter's demands, the SLN plans not only to continue Thio operations after its centennial anniversary but to open up new sites. Different from Thio, these new sites will be integrated into the Kanak environment. "The paternalism of the past is over" declared Philippe Gros, general director of the SLN.[90]

The SLN has also made efforts at reforestation and pollution control. The company has built dikes to channel the coarser material and a diversion dam was built to trap the finer ones to reduce the amount of material transported beyond the Wellington waterfall on one of the tributaries of the DoThio.[91] The difficulty lies with abandoned mines which continue to pollute the environment. It is the nature of the boom and bust cycle of the industry that new mines will open, only to be abandoned shortly after when the rush is over. It is these abandoned sites which pose the greatest threat to the physical environment of Thio:

> In the 1970s a territorial commission for the prevention of damage by mining was established in New Caledonia composed of representatives of public authorities, local agencies (municipal and Melanesian) mining companies and scientific organizations. Thereafter new mining necessitated an authorization from the Service des Mines following consultation with the commission which is required to make a safety inspection of the area to be mined, assessing the risks of erosion on and around the site and compiling

a botanical inventory to check if there are rare species or communities requiring site preservation. If mining is authorized the mining company is obliged to put in conservation works to prevent damage and pollution downslope an downstream from the mining areas. These include slope terracing behind boulder walls to achieve soil and waste mantle stabilization, barrages to intercept downwashed material and management to maintain percolation and prevent erosion along prospecting trackways and access roads. These procedures have been helpful but have been introduced far too late to prevent extensive damage. In particular, the many mines developed during the period of rapid expansion (1968-1973) and then abandoned are responsible for widespread and still uncontrolled erosion, waste dispersal down slopes and river pollution and aggradation, the effects of which are still spreading and will persist for many decades possibly centuries.[92]

Conclusions

The impacts of opencast nickel mining on the social and physical landscape of New Caledonia are likely to persist for a prolonged period. In terms of the environment, some improvement has already been observed. Government action has forced mining companies to exercise more care in their operations.[93] However, New Caledonia is remote from centers of conservation activism and the opencast mining has attracted much less criticism than if it were in France.

The omnipresence of nickel in Thio and the threat it represents to the locality by undermining other productive activities and limiting agriculture because of environmental degradation, reminds us of the necessity to coordinate activities on a national and regional level. Nickel, even if the market remains favorable will not create enough jobs to meet future employment demands in the community. If the community is to prevent an exodus and preserve its heritage, emphasis must be put on developing the agricultural sector. Conditions must be improved, that is, flooding must be curbed and agricultural incomes improved. This would mean coordination on a national scale so that agricultural activities become more important to the territorial economy. According to Tissier, this would imply new land ownership policies, new credit programs, and particularly new pricing policies concerning agricultural products so that agricultural activities can compete effectively with mining wages.[94]

Moreover, this needs to be coordinated with efforts at promoting an agriculture which builds on local institutions. Indigenous knowledge is often lost in the process of incorporation into the market economy because it becomes less relevant to the new situation and because it is systematically devalued by the process of specialization that competitive market production involves. If it is to be successful, economic productive activity needs to be fostered within the cultural frameworks of Kanak society. Eco-development practices have to be designed which are more compatible with indigenous social systems and which embrace the world views of islanders, including their ways of organizing their understandings of their social and material environments.

An essential aspect to the use of natural and human resources is the inclusion of indigenous cultural knowledge into the development paradigm. Incorporating the environment into development involves incorporating the socially constructed environments of Kanak culture. The experiences of the Kanak people in managing their natural environment should be an essential element in a more relevant approach to development in the region. However, it is precisely this cultural aspect which is absent from many proposed solutions to the current development challenges facing New Caledonia today.

Eco-development in New Caledonia would allow for the exchange of ideas and integrates cultural approaches to development needs. It is neither an imposition of foreign cultural models of development, nor is it a refusal to integrate useful concepts and experiences from other societies. It implies exchange and cross fertilization and would involve a democratic dialogue between policy makers and local communities with an emphasis on new forms of development co-operation that promote participatory development using decentralized channels.

Notes

1. La Grande Terre has a land mass of 16,750 square kilometers and the other islands total 2353 square kilometers for an overall total of 19,103 square kilometers.

2. The term "Canaque" was introduced to New Caledonia by Polynesian sailors during the period of early contact with Europeans, and in the local context it did have a pejorative meaning similar to that of "nigger" in North America. In the early 1970's the native peoples of New Caledonia changed the spelling to "Kanak" and this marked the birth of a black power type consciousness (see Jean Chesneaux, "Kanak political culture and French political practice: Some background reflections on the New Caledonian crisis," in M. Spencer, A. Ward and J. Connell, eds. *New Caledonia :Essays in Nationalism and Dependency*. Australia: University of Queensland Press, 1988). For the sake of clarity I have changed the word "Canaque" in the quotations to Kanak whenever suitable.

3. Each province has an elected Assembly, which is responsible for local economic development, land reform and cultural affairs. Together, the members of the Assemblies constitute the territorial Congress, which is responsible for the

territorial budget and fiscal affairs, infrastructure and primary education. Local government is conducted by thirty-two municipalities.

4. Because of labor migration to the territory in the post-World War II era, the Kanaks, have now become a minority in their own land. The current demographic profile of the territory is as follows: on 4 April 1989, there was 164,173 inhabitants. The territory's population has grown sixty-three percent in twenty years but this evolution has been very irregular because of the nickel boom in 1976. The Kanaks are the largest ethnic group with 73,598 people or 44.8 percent of the total. Nearly 34,000 Kanaks are under eighteen years old. The Europeans with 55,085 (33.6 percent) can be broken down into several groups: Caledonians born in the territory (known as Caldoche), immigrants from the 1960s and 1970s, metropolitan French living in the territory for a limited contract, and retired people. The Caldoches are about 30,000 or eighteen percent of the territory's total population. Seventy-eight out of one hundred inhabitants were born in New Caledonia. The vast majority of migrants to the territory are French citizens which reinforces the territory's ties to the metropole.

5. Following the discovery of significant nickel deposits in 1873, a nickel rush occurred between 1874 and 1877. Finding few other areas of economic growth, mining began to assume a much more important place in the colonial economy and some 600,000 tons of nickel were mined in New Caledonia between 1873 and 1900. As the industry continued to grow, between 1900 and 1923, New Caledonian nickel production amounted to over three million tons. In 1920, nickel and chromite accounted for forty-seven percent of New Caledonia's exports by value, but the depression of the early 1930s caused a severe crisis in the mining sector as demand and prices dropped sharply. The immediate impact of World War II on the colony was to seriously disrupt the New Caledonian economy. However, the arrival of American forces led to an abrupt change, producing an unprecedented economic boom. Among those who profited most from the war during the American occupation were the mining companies since export duties were suspended and prices for nickel and chromite soared. Demand and the price of nickel increased sharply in 1952, after the outbreak of the Korean War and the nickel market remained relatively healthy for the next ten years. The industry entered a short recession in 1962, until the New Caledonian economy experienced a nickel-based boom between 1969 and 1972 that had profound implications throughout the colony. Even before the boom began, nickel accounted for about ninety-eight percent of the value of New Caledonia's exports. The nickel boom came to an abrupt end in 1972, with a downturn in the world market in the face of oversupply and declining demand. The sudden end of the boom left New Caledonia in a critical position; see Michael Howard, *Mining, Politics and Development in the South Pacific* (Boulder: Westview Press, 1991), pp. 132-147.

6. Howard, *Mining, Politics and Development in the South Pacific*, p. 141.

7. Quoted in Michael Howard, *Mining, Politics and Development in the South Pacific*, pp. 141.

8. Quoted in S. Henningham, "France and the South Pacific in the 1980s: An Australian perspective," in *Journal de la Société des Océanistes*, Vol. 92-93, 1991, p. 22.

9. Nickel is a double edged sword for the future of the territory. It is a resource of considerable potential but one which assures continued dependence if not on the French state then certainly on the international world market. The nickel price is established by market demand and speculation in which trading on the London Metal Exchange plays a central role. The world nickel market is very unstable and is characterized by extreme fluctuations due to overproduction and overstocking of the metal. Another problem is that New Caledonian nickel is produced in a free zone but sold in American dollars on the international market and since one of the territory's buyers is Japan the difference between the yen and the dollar can have a serious affect on the industry. For example, in 1987 the SLN lost ninety million $US between declining exports and the slide in the $US compared to the yen.

10. M. Benezit, *Report on Mining Pollution in New Caledonia*, South Pacific Regional Environment Programme, Topic Review 01 (Nouméa, New Caledonia: South Pacific Commission, 1981), p. 1; J.F. Dupon, *The Effects of Mining on the Environment of High Islands: a Case Study of Nickel Mining in New Caledonia* (Noumea: South Pacific Commission, 1986), pp. 2, 4.

11. Benezit, *Report on Mining Pollution in New Caledonia*, pp. 1, 4, 6.

12. Dupon, *The Effects of Mining on the Environment of High Islands*, p. 4.

13. E. Bird, J-P. Dubois, and J. Iltis, *The Impacts of Opencast Mining on the Rivers and Coasts of New Caledonia* (Tokyo: United Nations University Press, 1984), p. 49.

14. Dupon, *The Effects of Mining on the Environment of High Islands*, p. 1.

15. J-P. Doumenge, *Du Terroir à la Ville.: Les Mélanésiens et Leurs Espaces en Nouvelle Calédonie*, Travaux et Documents de Géographie Tropicale No 46 (Paris: Centre Nationale de la Recherche Scientifique, 1982), p. 22.

16. The Mine du Plateau and its extension (Mine des Carrières), located between the Thio river in the south and the lower course of the Dothio river in the north, form the core of this complex. These two mines alone supplied seventeen percent of the total amount of ore produced since the end of last century, without a single break in production since 1935 (Dupon, *The Effects of Mining on the Environment of High Islands*, p. 3).

17. Bird, Dubois, and Iltis, *The Impacts of Opencast Mining on the Rivers and Coasts of New Caledonia*, pp. 14, 37; J. Tissier, *Les Bases du Développement Économique dans la Commune de Thio* (Paris: Institut de Recherches et d'Applications des Méthodes de Développement, 1989), pp. 4, 7.

18. Dupon, *The Effects of Mining on the Environment of High Islands*, p. 3.

19. J. Connell, *New Caledonia or Kanaky?*, Pacific Research Monograph No. 16 (Canberra: National Centre for Development Studies, Australian National University, 1987), p. 128.

20. Dupon, *The Effects of Mining on the Environment of High Islands*, p. 3.

21. Letter from Father Ameline, missionary in Thio, to Archbishop Monseigneur Fraysse, 30 August 1890 (extracts). Source: AAN Box 79, Archives de l'Archevêché de Nouméa.

22. Decree of 24 May 1889.

23. J. Dauphiné, *Les Spoliations Foncières en Nouvelle-Calédonie (1853-1913)* (Paris: L'Harmattan, 1989), p. 154.

24. Anonymous, "La société 'Le Nickel' de sa fondation à la fin de la deüxième guerre mondiale. 1880-1945," *Journal de la Société des Océanistes*, XI (11), 1955, pp. 99, 101.

25. Dauphiné, *Les Spoilations Foncières en Nouvelle-Calédonie*, p. 154. Land problems in Thio are still not resolved. For example, of the 100,000 hectares in the district of Thio, 85,000 belong to the French state, 12,000 to Europeans and most of this to three of them, while 1,700 Kanaks live on 3,000 hectares; C. Gabriel and V. Kermel, *Nouvelle-Calédonie: La Révolte Kanake* (Paris: La Breche, 1985), p. 189.

26. Letter from Father Dumussy, missionary in Thio, to Archbishop Monseigneur Fraysse, 7 February 1892 (extracts). Source: AAN Box 79 Archives de l'Archevêché de Nouméa.

27. Dauphiné, *Les Spoilations Foncières en Nouvelle-Calédonie*, p. 214.

28. Dauphiné, *Les Spoilations Foncières en Nouvelle-Calédonie*, p. 276.

29. Letter from Father Dumussy, missionary in Thio, to Archbishop Monseigneur Chaurion, 11 August 1913 (extracts). Source: AAN Box 79 Archives de l'Archevêché de Nouméa.

30. Tissier, *Les Bases du Développement Économique dans la Commune de Thio*, pp. 40, 43.

31. Letter from Father Ameline, missionary in Thio, to Archbishop Monseigneur Fraysse, 12 November 1890 (extracts). Source: AAN Box 79 Archives de l'Archevêché de Nouméa.

32. In 1901 Thio was the largest center having produced 42,000 tons of nickel; B. Antheaume, "Le Nickel et sa metallurgie," in *Atlas de la Nouvelle-Calédonie* (Paris: ORSTOM, 1981), plate 19.

33. Anonymous, "La société 'Le Nickel,'" pp. 99, 104.

34. Anonymous, "La société 'Le Nickel,'" pp. 99, 105.

35. Antheaume, "Le Nickel et sa metallurgie," plate 19.

36. For details see D. Winslow, "Workers in colonial New Caledonia to 1945," in J. Leckie, C. Moore, and D. Munro, eds., *Labour in the South Pacific* (Townsville: James Cook University Press, 1990).

37. See Dauphiné, *Les Spoilations Foncières en Nouvelle-Calédonie*, p. 253.

38. Bird, Dubois, and Iltis, *The Impacts of Opencast Mining on the Rivers and Coasts of New Caledonia*, p. 38; Tissier, *Les Bases du Développement Économique dans la Commune de Thio*, p. 29.

39. Tissier, *Les Bases du Développement Économique dans la Commune de Thio*, p. 26.

40. Benezit, *Report on Mining Pollution in New Caledonia*, p. 4.

41. Anonymous, "La société 'Le Nickel,'" p. 101.

42. Letter from Father Dumussy, missionary in Thio, to Archbishop Monseigneur Chaurion, 15 February 1919. Source: AAN Box 79 Archives de l'Archevêché de Nouméa.

43. Anonymous, "La société 'Le Nickel,'" p. 105.

44. "Histoire de la fondation de la mission de Thio," in *Bota Mère*, No. 21, 12 February 1908 (extracts), p. 15. Source: AAN Box 79 Archives de l'Archevêché de Nouméa.

45. Bird, Dubois, and Iltis, *The Impacts of Opencast Mining on the Rivers and Coasts of New Caledonia*, p. 18.

46. Letter from Anonymous Visiting Priest to Thio, in *Bota Mère*, No. 21, 16 October 1913 (extracts), p. 14. Source: AAN Box 79 Archives de l'Archevêché de Nouméa.

47. Letters from Father Dumussy, missionary in Thio, 11 March 1913, 16 March 1913, 29 December 1913. Source: AAN Box 79 Archives de l'Archevêché de Nouméa.

48. "Histoire de la fondation de la mission de Thio," p. 19.

49. R. Aldrich, *The French Presence in the South Pacific 1842-1940* (Honolulu: University of Hawai'i Press, 1990), p. 118, 290.

50. Howard, *Mining, Politics and Development in the South Pacific*, p. 136.

51. Anonymous, "La société 'Le Nickel,'" p. 110.

52. Anonymous, "La société 'Le Nickel,'" p. 111.

53. Anonymous, "La société 'Le Nickel,'" p. 113, 115.

54. V. Thompson and R. Adloff, *The French Pacific Islands: French Polynesia and New Caledonia* (Berkely: University of California Press, 1971), p. 403.

55. Tissier, *Les Bases du Développement Économique dans la Commune de Thio*, p. 7.

56. Dupon, *The Effects of Mining on the Environment of High Islands*, p. 4.

57. Dupon, *The Effects of Mining on the Environment of High Islands*, p. 3; Bird, Dubois, and Iltis, *The Impacts of Opencast Mining on the Rivers and Coasts of New Caledonia*, p. 38, 40.

58. Tissier, *Les Bases du Développement Économique dans la Commune de Thio*, p. 33.

59. Dupon, *The Effects of Mining on the Environment of High Islands*, p. 4.

60. Bird, Dubois, and Iltis, *The Impacts of Opencast Mining on the Rivers and Coasts of New Caledonia*, p. 40.

61. J-P. Doumenge, *Paysans Mélanésiens en Pays Canala*, Travaux et Documents de Géographie Tropicale No 17 (Paris: Centre Nationale de la Recherche Scientifique, 1975), pp. 97, 172.

62. According to J. Tissier (*Les Bases du Développement Économique dans la Commune de Thio*, p. 10), the principal territorial route links Thio on the west coast not to its surrounding communes but to Noumea, the territorial capital on the east coast and center of smelting activities, one hour forty-five minutes distance by car. Therefore Thio, with mainly a non-Kanak population, has been traditionally oriented to Noumea and the Doniambo smelter which absorbs almost of the commune's nickel production.

63. Tissier, *Les Bases du Développement Économique dans la Commune de Thio*, p. 23.

64. Tissier, *Les Bases du Développement Économique dans la Commune de Thio*, p. 25.

65. Doumenge, *Paysans Mélanésiens en Pays Canala*, p. 168.; Tissier, *Les Bases du Développement Économique dans la Commune de Thio*, pp. 24, 43.

66. Tissier, *Les Bases du Développement Économique dans la Commune de Thio*, p. 19.

67. Tissier, *Les Bases du Développement Économique dans la Commune de Thio*, p. 32.

68. Tissier, *Les Bases du Développement Économique dans la Commune de Thio*, p. 19.

69. Thio elder quoted in J. Tissier, *Les Bases du Développement Économique dans la Commune de Thio*, p. 24.

70. Doumenge, *Du Terroir à la Ville.*, p. 392.

71. Doumenge, *Paysans Mélanésiens en Pays Canala*, p. 110; Doumenge, *Du terroir à la ville.*, p. 374.

72. Doumenge, *Paysans Mélanésiens en Pays Canala*, p. 189.

73. Doumenge, *Paysans Mélanésiens en Pays Canala*, p. 171.

74. Doumenge, *Paysans Mélanésiens en Pays Canala*, p. 171.

75. Letter from Father Ameline, missionary in Thio, to Archbishop Monseigneur Fraysse, 9 November 1890 (extracts). Source: AAN Box 79 Archives de l'Archevêché de Nouméa.

76. Doumenge, *Paysans Mélanésiens en Pays Canala*, p. 171.

77. Father Rougé, Thio Mission Diary, 2 March 1949. Source: AAN Box 80 Archives de l'Archevêché de Nouméa.

78. Doumenge, *Paysans Mélanésiens en Pays Canala*, p. 172.

79. Howard, *Mining, Politics and Development in the South Pacific*, p. 157.

80. Howard, *Mining, Politics and Development in the South Pacific*, p. 156; Henningham, *France and the South Pacific*, p. 83.

81. See Gabriel and Kermel, *Nouvelle-Calédonie*, pp. 188-191 for details.

82. According to Gabriel and Kermel, (*Nouvelle-Calédonie*, p. 189), the mayor of Thio at that time, Roger Galliot, was a confirmed anti-independantiste, and a perfect symbol of the colonial bourgeoisie. He was a large land owner in the region of la Foa, boss of a nickel mine, stock holder in a fishing company, and one of the wealthy elite of the territory. His political career was also interesting, following the political evolution to the right of the holders of economic power as they began to realize Kanak intentions in the territory. He began as a member of the Union Calédonienne a moderate party promoting racial cooperation but left the party once it began to support Kanak land claims. After a brief time in a progiscardien party in 1979, he organized the PNC (Parti national caledonien) the New Caledonien national front party—an extreme organization of the right tied to the national party of Mr. Le Pen in France.

83. Gabriel and Kermel, *Nouvelle-Calédonie*, p. 52.

84. Howard, *Mining, Politics and Development in the South Pacific*, p. 156.

85. Marie-France Machero, quoted in *Le Monde*, 5 March 1985.

86. Gabriel and Kermel, *Nouvelle-Calédonie*, p. 59.

87. Tissier, *Les Bases du Développement Économique dans la Commune de Thio*, p. 25.

88. Howard, *Mining, Politics and Development in the South Pacific*, p. 157.

89. Rassemblement pour la Calédonie dans la République—the settler-dominated conservative political party which favors retaining a relationship with France.

90. Quoted in F. Bobin, "La mine néo-calédonienne en pleine recomposition," in *Le Monde*, 5 July 1991, p. 17.

91. Dupon, *The Effects of Mining on the Environment of High Islands*, p. 5.

92. Bird, Dubois, and Iltis, *The Impacts of Opencast Mining on the Rivers and Coasts of New Caledonia*, p. 51.

93. Powers are shared between the government and the judiciary and the territorial legislation reflects this shared responsibility by making separate provisions for two types of cases of mining damage: (1) in the case of prevention or reparation of damage the government (dept. of mines) shall subject the miner to control and surveillance (technical and administrative monitoring) and if necessary, shall take action itself. However, for this action to be legal, the damage must affect of threaten public security, conservation of water resources or public thoroughfares; and (2) in the case of repairing damage caused by mining activity to a third party or neighboring property and in the absence of an amicable arrangement between the parties, the party which considers itself injured shall have recourse to a court of law and may, if the court so orders, receive compensation corresponding to the value of the losses incurred (Benezit, *Report on Mining Pollution in New Caledonia*, p. 2). See also the article by Réné Sintes in *Les Nouvelles Calédoniennes*, 30 May 1991.

94. Tissier, *Les Bases du Développement Économique dans la Commune de Thio*, p. 67.

6

Smallholder Commercial Cultivation
and the Environment:
Rubber in Southeast Asia

Alec Gordon

In this chapter I place commercial rubber cultivation in the setting of the destruction of primary tropical forest and types of possible reforestation. The discussion will begin to deal with the environmental consequences of reforestation via the environmentally benign cultivation of the natural rubber tree (*Hevea brasiliensis*). The choice is deliberate and useful. I am able to utilize the paradox that although rubber trees may be regarded as an environmentally friendly commercial crop—they are after all rainforest trees themselves—all the large areas of rubber gardens in Southeast Asia (and elsewhere) were initially created by the destruction of original rainforest over the past ninety years. What is more, the tree is not even native to the region, having been imported from Brazil via Kew Gardens. Its career as a profitable and strategic crop has produced a sizeable modern documentation which permits a history of its commercial life.

What I propose to do is to accept provisionally the credentials of rubber as a reasonably sound environmental proposition then examine the actual conditions of its cultivation and the behavior of people in this production process to understand whether the specifics of the situation justify rubber's claim to environmental friendliness. In fact, rubber growers make no such claim! Their behavior has no subjective attachment to environmental criteria. They plant rubber trees for a livelihood. And this fact forms the basis of my approach. I do not propose to recommend for or against rubber cultivation on ecological grounds. I take an existing and relevant situation and study the expanding cultivation and processing of rubber by millions of small growers to determine its impact.

So, for the moment, let rubber cultivation be acceptable environmentally. In making this temporary assumption I have certain favorable evidence. First, it is a tree growing to twenty-five meters, a rainforest tree with a commercial life around thirty years. Thus, it seems a better primary forest replacement than, say, cassava, dry rice, maize, or sugar. There is also the positive subjective evidence of my own experience. On more secure ground, I can cite the good credentials given it by a number of authors.[1]

First, then, the setting of rubber cultivation—its extent, its value, its human culture, and so forth. This involves us having to stress the fact that rubber is overwhelmingly grown by small farmers and not by estates. The matter is crucial since it will be argued that the cultivation practices of smallholders are environmentally more suitable than those of the estates. Secondly, I shall outline in detail the smallholders' work process. Thirdly, I will match these processes to environmental standards and assess them. Finally, I will consider present trends and future possibilities.

The Setting

First time readers of the rubber trade literature are in for a shock. They will discover that natural rubber grows on trees, without, apparently, any human intervention. Alas, this is an illusion, caused by the excellence of data on production, prices, yields, high yielding varieties, fertilizer application, and tapping systems. Information about people is meager (with the partial exception of estates) and must be sought out. Luckily, of the information that does exist, some is very useful.

In summary, rubber is the third or fourth largest agricultural export of the Third World. About eighty-five percent comes from Southeast Asia, where the main concentration lies in Indonesia, Malaysia, and Thailand. Rapidly expanding smallholder production in Thailand over the past decade turned that country into the largest rubber exporter in 1991. In the region, smallholders account for three-quarters of the area and about ninety percent of a workforce, which is estimated at around six million people. At least half the workforce is composed of women, although most writing ignores this fact. The average size for estates is something over 300 hectares, while the average size of the smallholding is just over two hectares. Details are tabulated below with some comments.

The estimated world-wide production is shown in Table 6.1. Smallholder production has been growing steadily and in recent years has been more than double that from estates. In Malaysia output has been declining as rubber estates have converted to oil palm cultivation, while production in Thailand (mainly from smallholdings) has more than doubled (see Table 6.2). The accuracy of the statistics for smallholder areas may be doubted, but even so their predominance is clear. To indicate a fairly typical distribution of smallholders by size of holding, estimates for Thailand are shown in Table 6.3. The small size is very apparent. In the absence of reliable and consistent official statistics of the numbers of rubber smallholders, I made my own estimates in a project funded by FAO.[2] As later slightly amended following further research, these figures are shown in Table 6.4.

TABLE 6.1 Natural Rubber Production (thousand tons)

	1980	1986	1989	1990
Estates	1,490	1,565	1,610	1,550
Smallholdings	2,765	2,885	3,500	3,455
World Total	4,255	4,450	5,110	5,005
of which:				
Malaysia	1,530	1,539	1,419	
Indonesia	1,020	1,049	1,256	
Thailand	501	782	1,178	

Source: International Rubber Study Group, 1991.

TABLE 6.2 Estimated Areas Under Rubber in 1989 for Large Producing Countries

	Smallholders ('000 ha.)	Estates
Malaysia (peninsular)	1,488	369
Indonesia	2,589	522
Thailand	1,678	69
India	354	76
Sri Lanka	112	87
Brazil	138	59

Source: International Rubber Study Group, 1991.

TABLE 6.3 Distribution of Rubber Smallholdings by Size in Thailand, 1979

Size of Holding (hectares) hectares	Avg Size of Rubber Land (hectares)	Rubber Holdings, (% area)	Rubber (no. as a %)
0-3.18	1.1	20	48
3.19-6.38	2.6	34	34
6.39-9.58	3.9	18	10
9.59-22.38	8.1	17	5
22.39-39.99	24.0	7	3
40 and over	45.2	5	0.5
Average 4.48	2.66	Total 100	100

Original state areas in rai (1ha =- 6.25 rai)
Sources: National Statistics Office, 1981; Office of the Rubber Replanting Fund, 1979.

TABLE 6.4 Estimated Numbers of People Employed in Rubber Growing, for Three Largest Producers, 1986

	Employed Total	Of Whom Women
Malaysia	1,495,000	758,000
Indonesia	2,200,000	1,070,000
Thailand	1,100,000	550,000
Total	4,795,000	2,378,000
Of which in estates		
Malaysia	135,000	87,000
Thailand	55,000	28,000
Indonesia	200,000	70,000
Sub total estates	390,000	176,000

Source: Gordon, 1988, p. 72.

Given the facts concerning the extent of smallholder cultivation, it is indeed surprising to find in the work of a prominent ecologist such as Myers a belief that estates (or plantations) predominate and that their industrial practices are ecologically sound.[3] This error would be unimportant only if estate cultivation practices were similar to those of smallholders. As we shall see, they are not. Myers is, in effect, throwing his case away. It can be restored, however, and significantly improved, by looking more closely at the role of smallholders.

Rubber Cultivation Practices

All rubber cultivation practices, whether estate or smallholder, are "modern." As the crop was commercially grown in Southeast Asia before 1900, "traditional" techniques of cultivation did not exist to be used. Neither could the existing production methods already utilized in rubber's homeland, Brazil and Peru, be introduced. South American rubber was produced commercially from vast numbers of wild (not cultivated) trees that were tapped by adapting, intensifying, and debasing long established jungle collecting practices of the indigenous peoples. Rubber in South America was not grown: it was collected. Eventually it fell under such commercial pressure from Brazilian rubber dealers and overseas demand that the process became one of slaughter tapping. The cuts made in the bark of the rubber trees were made so frequently, so repeatedly, and so deeply that the wild trees were being killed in the process. The results of such practices were not dissimilar to forest destruction by logging. In addition, collecting of rubber cheaply on this basis was made possible by the brutal enslavement of local Indian peoples.

As there were no native stands of *hevea brasiliensis* in Southeast Asia to be worked, cultivation had to be undertaken. Planting and growing the tree in the region was relatively easy. It is not demanding as regards soil and will grow easily on level ground or on slopes, provided drainage is tolerable. Its rainfall requirements (1,400mm to 2,500mm a year) are met over extensive localities in the region. It tolerates a dry season of two months or more, although during this time the yield then drops off and tapping may become uneconomic. An elevation of 400 meters is regarded as its maximum, but it is met with at higher levels in the region.[4]

The actual "harvesting" methods had to be developed. Adapting American practices, pioneers originally gouged deep chevron trenches out of trees so that the liquid rubber (latex) might emerge and flow to collecting jars on the plant. This kind of tree mutilation was then recommended as possible only once every two or three years. This was scarcely viable economically and came to be replaced by better tapping methods. It was found that a small and shallow spiral cut on part of the tree stimulated an adequate flow of latex. Newer, adjacent (usually lower) cuts could be made successively on the trunk without undue removal of bark. After "resting," the area of the original cut can be tapped again. Thus, a tree could be tapped every three days (or even daily on smallholdings) and have a commercial life of about a quarter of a century after reaching maturity at seven years.

Initially, smallholder and plantation cultivation practices were not so far apart, but there were important differences. Thus, smallholders planted (and usually still plant) more trees for a given area. Also, smallholders were usually more skilled and experienced in tropical farming than were the plantation managers and their staff. These differences meant that the yield per hectare on Malaysian smallholdings exceeded that on estates. In addition, the estate mania for neat and tidy weeding between orderly rows of rubber trees created unnecessary erosion. Whereas, smallholder ground cover was maximized by their minimal or non-existent "weeding." Something of this contrast still holds.

As rubber cultivation became increasingly scientific, however, high yielding varieties were introduced which were generally responsive to increasing inputs of capital. As a result, the estates' productivity per hectare and per person outstripped that on smallholdings. However, it would be a mistake to regard the falling behind of smallholder productivity as solely related to such technical changes. Colonial policy, from the outset in the case of British Malaya and from the late 1920s in the case of Dutch Indonesia, opposed peasant growing in favor of the plantations. As noted by Lim, "the early history of peasant rubber cultivation is thus a story of a struggle against great odds."[5] Indeed, eventually the provisions of the British-run International Rubber Regulation Scheme would have wiped out peasant production altogether. As a report commissioned by the British Colonial Office pointed out, "the replanting of the estate acreage with high yielding material together with the prohibition of new planting would in the long run have eliminated the smallholders."[6] The attempt to put an end to peasant production failed. In part, the colonial regimes did not sufficiently control small farmers. Also, the intervention of World War II and the nationalism and revolution in its wake brought about a different power structure. Nevertheless, despite the achievement of independence, the technological bias against the smallholder, already in existence in all the restriction projects, remained for the future.

Rubber Smallholder Production Practices

As an essential introduction, I describe below how smallholder rubber is made and the respective roles of men and women (with some variations). Essentially, it is a family operation in order to encompass the range of tasks involved and which have to be completed within one day. This applies not only to family farmers but also to hired labor—larger farms in Thailand employ (and pay) sharecroppers as family units.

First, a comment on technology. Behind the trees now being tapped lies very sophisticated biotechnology. Mechanization is conspicuously absent (with the exception of transport), however, in the actual production of natural rubber. Biotechnology has increased the yield of the trees markedly—without, by and large, affecting the quality of the rubber. The quality of the product is determined in the harvesting and primary processing stages outlined below as well as in secondary processing.

Technology appears to have developed unevenly in natural rubber production. It is very high and still developing rapidly in terms of plant biology and is also high in the final levels of natural rubber processing. Many different types of high yielding trees have been in commercial production for thirty years or more and were in smallholdings. However, like most high yielding new varieties of plants, they require more intensive care than the older ones—notably in the greater application of fertilizer and pesticide. Furthermore, the HYVs, having been designed for estate cultivation, should be tapped only every third day or so. Generally, smallholders cannot carry out these practices. Their yields are, nevertheless, higher than before replanting with HYVs, although early tapping reduces the tree life drastically. In other words, the way in which plant biology high technology has been developed does not appear to have changed smallholder field practices significantly. As for standardizing the qualities or providing rubber with specified technical characteristics that has high value added, these steps are beyond the financial capabilities of all but some centralized smallholders. The same applies to improvements in smokehouse curing techniques. All those facilities are in the hands of estates, exporters, or government agencies.

Planting, although essential for production, is not detailed here. The tree life is thirty years or more so it is hardly a recurrent activity for the smallholder. During the six to seven years of maturing, the trees need and receive some attention, but the main significance of this waiting period for smallholders is that the family derives no income from the land and must have alternative activities or resources and/or receive credit. Temporary migration and/or sharecropping of rubber or other crops (such as pineapple) seem common responses.

Work for the smallholder begins very early. Before 1 a.m. is common. Carbide or battery "miner's lamps" carried on the forehead are used. The length of the working day varies according to the size of holding, number of people involved, and the distance to be traveled, but most of the activities are over by mid-day. This may permit other work to be done in daylight hours. A responsible person must be available to mind children during the parents' absence, but since the children will be asleep much of the time this is not so exacting for smallholders as it is for plantation families. The work is not continuous. Indeed, the latex takes time to flow and a further period elapses before it begins to solidify in the cup, so it is relatively easy to fit in other tasks, meals, or accompanying children to school, before it is collected.

Because of the small size of holdings and low productivity, the smallholders' need for ready cash implies that most will try to tap trees every day and sell sheet twice a week. However, heavy rain or the low yields in the dry season restrict the days that can actually be worked to about 150 or less in a year. This permits (indeed, necessitates) the undertaking of other economic activities by the rubber smallholders.

Tapping involves making a spiral cut (with a special knife) in the bark to cause the "bleeding" latex to run along the cut and drip into a cup. The latex may flow for several hours, during which time tapping other trees occurs, plus eating, resting, or childcare. Competence in tapping is easily acquired. Nevertheless, experts consider that women make better tappers than men as the task requires delicacy of touch so as not to damage the tree. (In estates, tapping is the highest paid field job.) However, this does not appear to be a prime factor in determining gender division of labor on smallholdings (nor on estates). In general, both men and women tap, but there are many cases of tapping by men only or women only. However, it is obviously very convenient for both to tap together and, having finished, return to the first trees to begin collecting latex.

Smallholder rubber may be produced in different ways. The most common method entails:

(1) *Collecting* latex involves emptying the cups into a bigger container, a can or a large churn, depending on the number of trees and yield. Children frequently assist. Transport may be simple. If the milling shed/hut is nearby, it is a matter of carrying a few cans a short distance. If the holding is too big or is far from a home-located shed, the latex may be transported in two churns by shoulder pole on foot, by cycle, or by motorcycle. Carrying latex churns is strenuous but is done by women at least as much as by men. Using a motorcycle may be more of a male preserve, but many women drive unisex motorcycles that permit skirts or sarongs to be worn with decency and in comfort. Even where males drive it is sometimes necessary for the women to ride pillion in order to hold the churns in position. Children assist here, too.

(2) *Coagulating* takes place in the milling shed, an open sided hut, that may be next to the home or in the rubber garden. The latex is sieved. Coagulant (formic, acetic, or sulphuric acid) is added and stirred carefully to remove air bubbles. Children help.

(3) *Milling* occurs after a time sufficient to let the latex coagulate. The slab of wet rubber is emptied from the dish onto the concrete floor and initially trod flat underfoot. (Less frequently, it may be flattened by hand rolling a metal tube over it. This strenuous task is normally done by young men.) It is then passed through two hand mills (costing around US$120

each in 1987 Thailand) to thin it, to wring out water, and to receive ribbing to assist further drying by air. The sheets will then be hung over poles, fences, etc. for two or three days to dry, usually in the open air but sometimes under raised houses. After that they are "unsmoked ribbed sheet" (USS) ready for selling. An individual sheet weighs around 1.2 kilograms. Before drying begins, and throughout the hand milling, much water should be used to clean the sheets. If not adequately done, mold will form which lowers the quality and price of the sheet. Moldy, discolored sheet is commonly seen. Water supplies can be a serious problem—for example, some smallholders have to carry buckets of water into more distant gardens. Within three to four days of tapping, smallholder rubber is thus ready for selling.

Despite the obvious presence of men, women, and children in coagulating and milling, there is what appears initially to be a sexual division of labor. This may not occur until the final stage of hand milling where the highest pressure ribbing mill is preferably turned by a man, the woman feeding and removing the sheet. This is "heavy work." But if the male is absent the women may "borrow" another male relative or simply do it herself. I have observed cases of women milling while young children fed the sheet.

Selling and marketing may occur at the farm, at a periodic market, or in town. The sheet may be sold to a resident local dealer, a traveling dealer, or to a smokehouse agent. To sell direct to a smokehouse the volume must be high so this is only done by large holders, by farmers' groups, or dealers. In Thailand, there seems to be no set gender division of this task, although it appears that women are more likely to sell small quantities and more frequently.

Smallholders also produce the following form of rubber in significant quantities: *cup lump*, where latex is left in the cup to coagulate and is collected on the next tapping day; *latex*, after being collected and sieved it is sold in liquid form; and *slab*. Mainly produced in Indonesia, slab rubber is formed by coagulating in large boxes or holes in the ground to form lumps of up to forty kilograms in weight. Cup lump and slab are usually inferior to and lower priced than unsmoked sheet.

In conclusion, it must be remembered that few rubber smallholders live exclusively on rubber. Most have other sources of income. The operation of a typical rubber estate may safely be left to the readers' own knowledge of large scale agribusiness practices. The general contrast between the two should be obvious. Specific comparisons are made below in the discussion of the environmental impact of smallholder rubber.

Environment versus Welfare?

It would be foolish to look favorably on the expansion of a crop area that did not confer some welfare benefits on its growers. And here certain paradoxes appear, should policy recommendations be considered. Most rubber smallholders have moderate incomes at best. In Malaysia, the incidence of poverty among rubber smallholders is officially considered to be high. On the other hand, in Indonesia and Thailand rubber growing is favorably regarded by officials and farmers alike. These differences reflect in part subjective judgments and in part levels of economic growth and income—Malaysia having a notably higher standard of living. At any rate, because the number of holdings and the total area they cover continue to grow, it would seem that many ordinary rural people regard rubber cultivation as a more desirable way of earning a living than other possibilities open to them. It may also be added that their income levels, moderate as they have been, were, nevertheless, subject to taxation burdens whose incidence rated among the highest in the world.[7]

What then might be done to improve the lot of the rubber smallholder and what is likely to be done? While these two questions are quite distinct and technical solutions (higher productivity) involve a different philosophy from social ones (land reform), the practical implications may be closer than imagined. Let us consider two different approaches as elaborated in Malaysia. (The issues have not been formulated with the same clarity elsewhere.) While the approaches begin far apart, the "social" one, as exemplified in the work of Shamsul Bahrin and Husin Ali,[8] and the "technocratic" one, as found in the report of the Expert Task Force,[9] converge, insofar as both recommend that the rubber smallholder stop being a small farmer. There are, however, important differences. Shamsul and Husin urge a general increase in the size of holding:

> Since shortage of land is the key factor in the plight of the smallholder, increased acreage of holdings becomes a solution. The strategy of alienating more land to needy smallholders must necessarily become a major exercise...Without an adequate holding all other solutions and improvements cannot be anything more than merely academic or cosmetic measures.[10]

In contrast, the Expert Task Force concentrates on "progressive" rubber smallholders whose land ownership and productivity is already above average. The experts' neo-classical viewpoint is expressed with admirable clarity and brevity:

It should be noted that the usual method of facilitating the adjustment of less progressive and failing farmers in a developing rural economy is for them to sell their land to their more progressive neighbour...using the financial proceeds to assist their transfer to other occupations...attention should be directed to facilitating land transfers.[11]

In other words, the Expert Task Force urges that the "normal" mutual process of growing landlessness among smaller farmers on the one hand, and acquisition of their land by larger farmers, landlords, corporations, and speculators on the other hand, should be encouraged. Through this euthanasia of the rubber smallholder, a much bigger, viable rubber holder of the future will emerge. This, unfortunately, is a more likely prospect than the broader increase in holding-size among smallholders advocated by Shamsul and Husin. Either way, a larger average rubber smallholding would emerge and on it cultivation practices would begin to resemble more closely those on the estates. Thus, from an environmental perspective, many of the practices on existing smallholdings that may be environmentally favorable could be lost. This need not be the case, of course. Were the research and development programs and extension services based on smallholder needs and practices, instead of the assumption that what is good for estates should work for small farmers as well, then an ecologically sound prospect would be in view. This would, however, be a utopian expectation. What then should we expect? And what are existing smallholder realities?

The Environmental Impact

Myers waxes quite lyrical about natural rubber trees as an ecologically desirable and valuable cultivated tropical forest.[12] He views it as a tropical forest that is now operating as a large "forest industrial complex" and with even greater potential.[13] He enthusiastically considers extending the range of rubber and allied trees to be cultivated both as a means of producing renewable hydrocarbon fuels and as a means of conserving something very close to an original rainforest. Because he is ignorant of the fact that most rubber areas are cultivated by smallholders, he weakens his case and actually makes recommendations which are not environmentally sound. Believing, as he does, that most rubber comes from estates, he seems to envisage merely an extension of their practices.[14] As will appear below, this seems an ambiguous proposition at best. Moreover, his assessment of the likelihood of a vast extension of rubber tree areas due to demand is premature. His naiveté in matters economic leads him to flounder into making forecasts that have already been discredited. Thus, he innocently takes World Bank prognostications at their face value (something the rubber trade does not do) and projects a near future (already past and disproven) of desperate rubber scarcity and rocketing prices.[15]

While noting these failings, his case may be accepted, which is founded on his own grounds of expertise, that commercial rubber cultivation may be considered environmentally desirable. This view is supported by scholars more familiar with the region and with the systems of cultivation. Donner argues that *hevea brasiliensis*, "a jungle tree which does not require great soil fertility, still seems one of the most useful plants in replacing the original jungle."[16] Likewise, Scholtz states that "the cultivation of perennials preferably rubber, would best meet ecological requirements" in the region.[17]

Bearing in mind the above-mentioned cultivation and processing practices, I now wish to consider their environmental impact under the headings of (1) erosion, (2) monoculture, (3) pollution, and (4) wildlife.

Erosion

On the face of it, there may seem to be little threat of erosion at all. The rubber tree has a long commercial life, twenty-five to thirty years in smallholdings, so felling and replanting does not occur frequently. Likewise, exposure to the elements is infrequent and brief. Nevertheless, the ground is bared after felling and it will be two or three years before the new rubber plants have much effect on holding the soil. The potential problem arises from the fact that rubber is generally grown on sloping ground and, indeed, frequently on hilltops (i.e., on watersheds). Where replanting occurs under official schemes, proper terracing of the rubber gardens is regularly undertaken (otherwise the replanting subsidy is withheld). In the extensive unsubsidized replantings, things may be less well-ordered but terracing seems to be quite common. Apart from anything else, it will make tapping easier once the trees are yielding. During the immature period, intercropping may be practiced, even if with the "wrong" intercrops (according to officialdom). In the less well-tended gardens, weeds (grass) are quite likely to predominate. Weeds do the rubber tree little good but they do prevent the ground from remaining bare. At any rate, one of the horrific features of new hill farming in Thailand, vertical ploughing of hillsides, is absent in smallholder rubber.

Problems do not end with the presence of an umbrella created by the tree canopy. Consequently, while the presence of vegetation does control erosion, different trees have canopies of different effectiveness. Quite how natural rubber rates in this is not entirely clear. In any case, it is frequently argued that the ground cover, not the canopy, is decisive in controlling erosion, since the rainwater has to hit the ground sometime. Surprisingly, the throughfall of water may have a higher erosive effect than direct rainfall. The real barrier to erosion at this point is the litter (fallen leaves, branches, etc.) on the ground. Donner has summarized some of the research findings for Indonesia (I again draw attention to the merging of different factors causing erosion that obscures the analysis somewhat):

> in the case of coffee, *coffea robusta* is less protective than *coffea arabica*. Rubber plantations are believed to give little protection because of the activities of the planters press the soil together so that the rainwater runs quickly away...generally rubber plantations are more prone to erosion than coffee gardens...The litter, however, is the more important anti-erosion agent.[18]

What all this appears to mean is that: (a) the rubber tree, though not the best, is better than no tree at all in providing a canopy; (b) because ground cover (litter) is more important than canopy cover as an anti-erosion agent, the crucial issue is what people do under the trees—in effect the intensity of weeding; (c) rubber estates, because they weed intensively, are only moderately good barriers to erosion; and (d) smallholders do not weed very much, so they maintain good ground cover (though it is a difficult work area at times). Thus, smallholder rubber should rate fairly highly in anti-erosion.

Presumably, it was considerations such as these that led Sholz to conclude of farming in large areas of Sumatra that "perennials, preferably rubber, would best meet ecological requirements."[19] An extreme form of ideal smallholder rubber garden might be one where long term swidden cultivation is practiced around and occasionally in a stand of rubber trees: "ecologically speaking, a swidden rubber forest is closer to the secondary forest than a rubber plantation."[20] However, the prospects for this declining form of cultivation are hardly promising. Still, other promising forms of multi-cropping are considered in the next section.

I conclude then by noting that smallholder rubber cultivation looks reasonably good against erosion, but estates are suspect.

Monoculture

For widely accepted reasons, I shall consider monoculture as environmentally undesirable. Equally simply, I can say that rubber estates practice monoculture while smallholders do not. Strictly speaking, there are no rubber smallholders. Rather there are small-scale farmers who cultivate rubber along with other crops (although rubber may be the most important). The other crops are usually wet or dry rice and/or fruit trees. Table 6.3 showed that in Thailand those classified as "rubber growers" had something like an average of 1.5 hectares of land in addition to nearly 2.5 hectares under rubber. Of course, the smallholders' land need not be in one plot. Indeed, it is common (actually necessary) in a village for all the rice plots to be in one spot and the rubber gardens together in a higher area. Although contiguous, however, these rubber plots never form the vast stands of rubber found in Malaysian estate regions. Many small growers actually have other species of tree growing inside the rubber plot and, depending how far "weeding" is "neglected," these look quite jungly.

Tree intercropping for smallholders has been grudgingly recognized as desirable by agro-economic authorities. For example, in Sri Lanka a model farm for a multi-species cropping system with multi-storey canopy arrangement has been set up. It is, in fact, a new version of the traditional highland forest garden of the Kandy region.[21] This model has tea as the main crop, but with modifications could apply to rubber. In fact, it has been in operation in many "rubber" smallholdings for decades as a version of the "Java Garden" long before the Java Garden became a vogue among enlightened agricultural economists and environmentalists. These diversified rubber plots are impressive to see and are very beautiful. However, the few I have personally viewed are operated by rather elderly people. I would guess that there is a risk of losing the very real skills needed to operate them through their not being passed down to a younger generation—at the very time that the glories of the Java Garden are at last gaining recognition.[22] However, a not dissimilar system has been successfully developed for rubber in China.[23]

Long Yiming and Zhang Jiahe have reported that several benefits were derived from rubber interplanted with tea and rubber interplanted with devil pepper in China's Yunnan Province, which they termed a natural plant community.[24] These benefits were: (1) substantial litterfall which improved soil fertility, (2) soil and water control due to the layering of canopy, (3) reduced stress from climate and other natural factors, and (4) greater economic returns compared with pure culture. This system mimics a natural and advanced seral stage.

The larger the smallholding the less diversified the plot is likely to be because of the greater concentration of rubber. But in size, even larger smallholdings cannot remotely compare with the monoculture of the larger estates.

Pollution

Various problems arise in relation to pollution. These include the disposal of acids used in primary processing, atmospheric pollution from secondary processing of the rubber into smoked sheet, and the side effects of chemicals (particularly pesticides) used in cultivation.

Unless steps are taken to prevent it, the concentration of rubber processing in estates gives rise to problems of acid disposal. This occurred in Malaysia and eventually provoked legislation requiring the treatment of waste. By and large, the legislation would seem to be enforced. The situation in Indonesia and Thailand is not known. However, the bulk of production is undertaken in smallholdings and there the results of disposing acid waste (after coagulation and washing) are not recorded. The most common disposal method appears from observation to be: (1) emptying a bucket just outside the milling shed, (2) for excess moisture to be milled out of the sheet and dribbled to the floor of the shed, and (3) the sheet to be splashed with clean water just outside the shed then hung to drip and dry on the garden fence or under the house. Clearly some acid waste must be entering the ground. What happens after that no one seems to know. Insofar as the naked eye may be trusted, no obvious adverse effects were observed. More certainly, the waste is dispersed into small parts as a result of the production units being small and scattered.

Air pollution through smoke is not caused by smallholders because they do not cure their own rubber. However, their unsmoked sheet (USS) has to be cured after sale by someone else—a smokehouse/exporter, estate processor, or government curing station. The old smokehouse method, which is still in use, is a very smoky business indeed, although the buildings are usually dispersed. However, as the point of the process is to keep the smoke inside with the rubber sheets, efficiency criteria developed efficient burners and effective emission control devices. On the outside it looks pretty well smoke free. At any rate, one does not see those huge columns of smoke by day and pillars of fire by night that typify palm oil refineries and disfigure the peninsular Malaysian and Sumatran countryside.

Whether the use of other chemicals raises serious problems is easily dealt with. Estates use them intensively and extensively. By and large, rubber smallholders do not use them so much. Therefore, the question is not so relevant to smallholders. Or at least not yet. Rubber growers in official replanting schemes are obliged to use fertilizer during the six or seven years of tree immaturity. At maturity the subsidy stops and, for small growers, so does the use of fertilizers. It should be noted that all the HYV trees are designed to be productive with the application of fertilizers so that the smallholders do not derive anything like optimum yields. Herbicides are used intensively by estates to control weeds and frequently with chemicals such as paraquat whose use is banned in most developed countries. Weeding on smallholdings is minimal.[25]

Those wishing to improve their living standards and able to do so are under constant pressure to act like estate managers. And with the tree technology on offer this is the way to increase yields. In other words, the small rubber farmer, who is at present fairly innocent of using pollution-risking chemicals, has to move in a potentially polluting direction to increase productivity.

Wildlife

There does not appear to be a great deal of material on this topic. At least, smallholders certainly avoid the excesses of monoculture. Although the tree seasonally flowers, rubber does not seem to produce much in the way of animal food. (Two migrant workers in east Java once told me that they preferred working in rubber gardens to coffee gardens because they could always climb a tall rubber tree to get out of the way of tigers!) Personal observation has shown that far fewer birds and animals are to be found in rubber gardens as compared to oil palm estates. But then the oil palm produces a fruit much sought after by insects, rodents, and birds who, in turn, attract predators—all living and some dying amidst heavy applications of pesticides. Visually attractive—one can see more than in a rubber garden because of extensive open ground occupied by the effluent treatment lagoons—the oil palm estates may be more dangerous to wildlife than the more neutral aspects of rubber. House gardens of rubber smallholdings often have quite varied bird life but, of course, this would be true of virtually any house garden. The standard work on Thailand's mammals remarks:

> Rubber gardens cover about 13,000 square kilometers in areas originally covered by evergreen rainforest or semi-evergreen forest. Whilst human activities deny rubber gardens to most larger mammals except in a very transitory form...[smaller] mammals...have all been found to inhabit rubber plots, feeding on leaves or insects. Some of these species may be permanent residents though most gardens are cleared and replanted every 40-60 years.[26]

The standard guide to birds comments:

> There are large areas of rubber and oil palm. Such tree crops have little wildlife importance supporting few, if any, native forest birds...Untended rubber gardens in which a moist understory of native forest plants has been allowed to regenerate will sometimes support a number of smaller more tolerant forest birds.[27]

At any rate, we may conclude that rubber gardens provide some kind of refuge for wildlife ranging from fairly minimal in busy estates to fairly good in unweeded smallholdings.

Conclusions

From the above discussion, it looks as though smallholder rubber production, as it is presently practiced, is quite a good environmental proposition. The previous socio-economic sections describe the polar extremes and unity of developments that are occurring in rubber-related land use and tenure whether environmentalists like it or not. Yet we are not doomed to be powerless. We are only impotent in our ignorance of the world of people and in our attempts to superimpose ideal environmentalist policies upon a recalcitrant socio-economic world. If I have demonstrated that smallholder rubber growing is environmentally desirable, we cannot fly to recommending its extension. Actually, environmentalists are here in no position to recommend anything because as yet there is no constituency to whom the recommendations can be made. It is true that in Malaysia some steps have been taken, but elsewhere rubber growers and environmentalists live on different planets. The problem is how to link these people politically while recognizing the socio-economic needs and aspirations of the rubber growers.

Although we live amidst a terrible threat of environmental destruction, this should not lead us into dreaming that we can mitigate it with authoritarian moralistic edicts. Thus, while I have approvingly quoted Donner above on relevant issues, the all too common approach that his desperation leads him to is self-defeating and must be rejected. For example, he says,

> if we consider the history of soil conservation in Indonesia and the experiments, discussions and recommendations produced over a century, we can only share the frustration of the few responsible specialists fighting against an ever-growing flood of millions of people who spread over the land, cut down the trees and till the ground, watching helplessly as it is washed away leaving the ground bare.[28]

This is typical of much of the literature, although it is unusually clear in the exposition of the attitude. It introduces a neo-technocratic, pseudo apolitical approach into the environmentalist movement. With due respect to his genuine indignation, we have to say, "no." We are not going to fight against a flood of millions of people. That would really spell disaster. If they are not potential allies then the game is lost. The "few responsible specialists" are but pawns whose ability can be transformed into effectiveness only by collaboration with "the flood."

Thus, I would not recommend forcing or even encouraging people to plant rubber trees for environmental reasons. Nor would I sing indiscriminate praise of smallholder farming. I am in no position to do so. However, it is encouraging surely to find a large commercial smallholder crop that is environmentally tolerable. Presumably, there must be more out there. A search would prove rewarding. Above all, an appreciation of why people in society do these things for reasons other than environmentalism and how they organize (and are organized) to do them is called for. By means like these, serious environmentalists can begin to establish a constituency. Then, making recommendations may be in order.

Notes

1. See N. Myers, *The Primary Source: Tropical Forests and Our Future* (New York: Norton, 1985); and W. Donner, *Land Use and Environment in Indonesia* (Honolulu: University of Hawaii Press, 1987).

2. A. Gordon and N. Sirisambhand, *The Situation of Women Rubber Smallholders in Southeast Asia* (Bangkok: Chulalongkorn University Social Research Institute, 1987), p. 18.

3. Myers, *The Primary Source*.

4. See C. Barlow, *The Natural Rubber Industry* (Kuala Lumpur: Oxford University Press, 1978); Rubber Research Centre, *Rainfall Patterns and the Planting of Rubber in Thailand* (Hat Yai: Rubber Research Centre, 1975).

5. Lim Teck-Ghee, *Peasants and Their Agricultural Economy in Colonial Malaya, 1874-1941* (Kuala Lumpur: Oxford University Press, 1977), p. 75.

6. P.T. Bauer, "The working of rubber regulation," in *Economic Journal*, September, 1946, p. 403.

7. See A. Booth, "Agricultural taxation: A survey of issues," in P. Shome, ed., *Fiscal Issues in Southeast Asia* (Singapore: Institute of Southeast Asian Studies, 1984); Khay-Jin Khoo, "The marketing of smallholder rubber," in Kamal Saleh, ed., *Rural Urban Transformation in Malaysia* (Tokyo: United Nations University, 1978). The taxes are levied at the port but the burden is transferred back to the primary producer.

8. Shamsul Bahrin and Husin Ali, "Challenges facing the smallholder," in *Ilmu Masyrakat*, no. 8, 1985, pp. 76-82.

9. Expert Task Force, *The Malaysian Natural Rubber Industry: Report of the Task Force of Experts* (Kuala Lumpur: Malaysian Rubber Research and Development Board, 1983).

10. Shamsul Bahrin and Husin Ali, "Challenges facing the smallholder," p. 82.

11. Expert Task Force, *The Malaysian Natural Rubber Industry*, p. 43, 78.

12. Myers, *The Primary Source*, p. 8.

13. Myers, *The Primary Source*, pp. 237-241.

14. Myers, *The Primary Source*, p. 239.

15. Myers, *The Primary Source*, p. 240.

16. Donner, *Land Use and Environment in Indonesia*, p. 13.

17. Scholtz as quoted in Donner, *Land Use and Environment in Indonesia*, p. 176.

18. Donner, *Land Use and Environment in Indonesia*, p. 132.

19. As quoted in Donner, *Land Use and Environment in Indonesia*, p. 176.

20. Donner, *Land Use and Environment in Indonesia*, p. 234.

21. J. Jacob, "Development of uneconomic tea and rubber lands," in W. Gooneratne and D. Wesumperuma, eds., *Plantation Agriculture in Sri Lanka.* (Bangkok: ILO/ARTEP, 1984), p. 202.

22. See, for example, World Resources Institute, "The Java garden," in *World Resources 1988-89* (New York: Basic Books, 1989), p. 59.

23. P. Sagise, "Plant succession and agroecosystem management," in A.T. Rambo and P. Sagise, eds., *An Introduction to Human Ecology Research on Agricultural Systems in Southeast Asia* (Los Baños: University of the Philippines Press, Los Baños, 1984), p. 147.

24. Long Yiming and Zhang Jiahe, as cited in A.T. Rambo, J. Dixon, and G. Wu, eds., *Ecosystem Models for Development*, Workshop Report (Honolulu: Environment and Policy Institute, East-West Center, 1984).

25. Rubber Research Institute, *Rubber Owners Manual* (Kuala Lumpur: Rubber Research Institute Malaysia, 1983), pp. 64-66, 266.

26. Boosong Lekagul and J.A. McNeely, *Mammals of Thailand* (Bangkok: Saha Karn Bhaet, 1988), p. xxviii.

27. Boosong Lekagul and P.D. Round, *A Guide to the Birds of Thailand* (Bangkok: Saha Karn Bhaet, 1991), p. 11.

28. Donner, *Land Use and Environment in Indonesia*, p. 123.

7

The Political Economy of Eucalyptus: Business, Bureaucracy, and the Thai Government

Apichai Puntasen, Somboon Siriprachai, and Chaiyuth Punyasavatsut

This chapter will seek an explanation of how eucalyptus has been introduced by the Royal Forest Department (RFD) of Thailand under the guise of a reforestation program. What are the hidden agendas that have never been explicitly specified by the RFD? Is the argument of reforestation only used to redirect attention from environmental questions? Are there any business interests involved in the introduction of the trees, and, if so, what is the significance of such involvement?

Eucalyptus is used here as a case study of economic interest that may explain the reaction of the Thai government and its bureaucracy in response to short-term politico-economic gain for those involved. Unfortunately, such reactions may bring about ecological and environmental damage and long-term economic disaster.

The Fast-growing Pulp and Paper Industries

After 1958, the Phibun regime and its economic nationalism was overthrown and replaced by the Sarit- and American-supported governments of Pot and Thanom. These new regimes propagated private business and foreign investment as a strong economic instrument to combat communism in Thailand. The pulp and paper industry was one of those encouraged and, since then, it has undergone rapid growth. While the annual growth rate of the industrial sector as a whole from 1974 to 1985 was 7.8 percent, the rate for the pulp and paper industry was 8.1 percent. During this period, the total amount of all kinds of paper consumed increased from 260,000 tons per year to 630,000 tons per year. Production of industrial paper increased from 97,000 tons per year to 294,100 tons per year, representing an annual growth rate during this period of 10.6 percent.[1] In 1985, industrial paper provided seventy-one percent of total demand for paper.

Data of pulp production before 1982 are rather sketchy. Most pulp used in paper production was imported. As a result of the 1973-74 oil price shock, the price of pulp in the world market increased very rapidly. This price rise induced the domestic industry to increase production. Pulp production statistics have been made available since 1977. According to these figures, production capacity increased from 39,000 tons in 1977 to 104,500 tons in 1982.[2]

Rapid increase in the demand for industrial paper and the rapid increase in imported pulp have resulted in considerable increases in the price of domestic pulp. Thus, the price increased from 11,576 baht per ton in 1986 to 13,169 baht per ton in 1987, and increased again to 15,391 baht per ton in 1988. The average rate of increase during the mid-1980s was 15.3 percent annually. Growing demand for pulp prompted many firms to apply for promotional privileges from the Board of Investment (BOI). In December 1988, the BOI reported that five pulp manufacturers had received such privileges.

In spite of its rapid expansion, domestic pulp production has not been able to keep pace with demand. Since 1986, most firms have been asked to produce pulp over their capacity. For example, in 1986, while total production capacity was 104,500 tons, actual pulp production was 116,732 tons. Yet, an additional 27,987 tons of pulp still had to be imported. In 1989, while domestic production increased to 146,894 tons, another 43,327 tons had to be imported. The figures indicate a considerable potential still for expansion of the pulp industry.

While the domestic demand is very strong, the supply of raw materials traditionally used in pulp production is decreasing at a very rapid rate. This situation has encouraged the search for alternative supplies of raw materials. Eucalyptus provides such an alternative and as soon as eucalyptus has become available, enterprises have rushed to apply for promotional privileges from the BOI.

Eucalyptus as a New Source of Raw Material

At the beginning of the First National Economic and Social Development Plan period, in 1961, the forest covered fifty-three percent of the land area of Thailand. The plan specified that fifty percent of the land area must be reserved for forest. Fourteen years later, in 1974, the forest area had been reduced to forty percent. Under the Fifth Plan (1982-86), in 1985, it was found that only twenty-eight percent of the land was left as forest, or ninety-three million rai. Of this amount, nine million rai, or ten percent of the remaining forest land, is the watershed area. The RFD became alerted to this situation and responded with a serious program of reforestation.

From 1965 to 1985, the total amount of reforestation covered 3,375,507 rai of land, while the average rate of deforestation was three million rai per year.[3] By 1985, it was apparent that if such a pattern continued, before long the remaining forest would disappear. When the Sixth Plan was formulated in 1985, a series of conferences were held to gain expert opinion for a national forest policy. The policy that was presented contained twenty points.[4] The five most relevant for the present discussion are as follows:

4. There should be at least forty percent of forest land for the two main purposes:

   4.1 The conserved forest will be used to conserve environment, soil, water, plant species, and rare animals, and to protect from natural catastrophes resulting from flood, land slides, including the use of the forest land for studies, research, and public recreation. The amount of conserved forest will cover fifteen percent of the total land area.

   4.2 Economic forest will be used for wood and forest products for economic purposes. The area covered will be twenty-five percent of the total land area.

5. The state as well as private enterprise shall develop the forest area in order to reach the said goal. The state will manage the forest in order to provide a constant flow of direct and indirect benefits to the general public.

13. The state shall support industries having linkage with the forest and the pulp mills in order to make use of all parts of woods and to encourage the use of wood substituting materials.

17. The state shall define the area with the average slope of more than thirty-five degrees to be forest area. The area will not be given any deed or title or certificate of land use according to the Land Act.

19. The state shall provide incentive for the promotion of silviculture by private enterprise.

Some of the above policy was translated into action plans in the Sixth Plan. According to this plan, the RFD would grow trees and improve the water shade area at the rate of 100,000 rai per year. It would grow forest in areas of high elevation at the rate of 300,000 rai per year. The RFD also proposed under this plan for private enterprise to be given forest concessions within national parks and for private enterprise to take part in the management of conserved forests, including the organization of hotels and tourist facilities within such areas.[5] This part of the plan met with strong public opposition and was, finally, abandoned.

By the late 1980s, there were an estimated 500,000 rai of eucalyptus in Thailand. Half of this area was privately owned. About 200,000 rai was grown in the northeast. Another 100,000 rai was grown in the north. The planted area in the central plain was 150,000 rai. There was about 50,000 rai of eucalyptus in the south.[6]

Although eucalyptus was first introduced to Thailand in 1946, large scale planting only began in 1978. It was found in the early 1970s that among the 600 species of eucalyptus in Australia, eucalyptus camaldulensis was the only variety that had the ability to grow successfully in a wide range of environments. However, the cost of the seeds of this species— 100,000 baht per kilogram—was sufficiently expensive to discourage promotion of the tree before 1978. But as soon as a cheap supply of the seeds was found in Australia (1,000 baht per kilogram), the planting of eucalyptus took off.

The availability of cheap seeds from Australia allowed for handsome profits to be made by eucalyptus nurseries from the sale of young trees. Each kilogram of seeds produces around 500,000 to 600,000 young plants, and each plant is sold to farmers and planters for between 0.80 and 1.50 baht. There appears to be a close association between some of the bureaucrats in the RFD (especially those in the Division of Private Forest Promotion) and their counterparts in the private sector who own nurseries or act as brokers. The bureaucrats in the RFD have strongly encouraged the public policy favoring growing eucalyptus camaldulensis all over Thailand by both the public and private sectors.[7] After 1978, during the Fourth Plan (1977-81), this group within the RFD was successful in lobbying for inclusion in the Fifth Plan a goal for the private sector to plant 300,000 rai of such eucalyptus per year According to the plan, the private sector would receive credit from commercial banks as well as related government agencies (such as the Bank for Agriculture and Cooperatives).

Five province in different parts of the country were identified as tree-growing centers where industries for wood processing would be established: Tak in the north, Khon Kaen and Surin in the northeast, and Cha-Choengsao and Rajburi in the east and south of the central plain. An area within a 200 kilometer radius of the industry was to receive special encouragement for planting fast-growing trees. Overwhelmed with enthusiasm, the then governor of Tak province, Kaj Rakmanee, started his policy of the "Green and Moisture of Tak" by encouraging the whole province to grow eucalyptus with the aim of supplying wood to the pulp industry that was to be established.[8]

Resulting from the Fifth Plan, on 11 December 1984, the cabinet made a resolution that 1985 to 1988 were to be declared National Forest Years. The steering committee of the National Forest Years recommended the growing of eucalyptus for commercial purposes as the "master plan" for the undertaking.[9] Government agencies, state enterprises, and adults and youths in all educational establishments were to be encouraged to plant fast-growing trees all over the country. Naturally, the government agency which propagated these activities was the RFD. It is also within this context that the national forest policy was formulated in 1985.

Woodchip can also serve as raw material for other industries, such as for the production of fibreboard and cement board. While many of those in the woodchip industry in Thailand have plans to develop domestic pulp mills, in order to reduce their risk, initially they focused on producing woodchip for export to Japan and Taiwan for further processing.

This export orientation is the primary reason for the concentration of woodchip factories in Cha-Choengsao. These plantations are in an ideal location for shipment abroad through Laem Chabang, a new deep-sea port. Yet, in terms of social and environmental costs, Cha-Choengsao is not the best location for eucalyptus plantations because of the fertility of the land which is more suitable for growing, for example, fruit trees or rubber trees.

Japanese and Taiwanese interest in Thai eucalyptus has been considerable. During 1987-88, representatives from the Paper Association of Japan wanted to form a joint venture in Thailand to grow eucalyptus and support the expansion of eucalyptus plantations, as well as to purchase unlimited amounts of eucalyptus from Thailand. The plan was rejected by Thai companies for fear that the proposed joint venture would create a monopoly and result in suppression of the purchase price. After the failure of this proposal, another joint venture was suggested in 1989, Thai Eucalyptus Resource, again, with the aim of purchasing unlimited supplies of eucalyptus woodchip.[10]

The largest companies in the woodchip and pulp industry generally are joint ventures with Taiwanese or Japanese interests. The largest of these is Suankitti Company Limited, a joint venture with investors from both Japan and Taiwan. It is involved in the eucalyptus plantation, pulp, and woodchip industries. The company has built Thailand's largest mill, with a higher production capacity than all other existing mills combined. After beginning operation in 1992, it was anticipated that the company would produce 330,000 tons of pulp. This amount exceeds domestic requirements and it is estimated that it will not be until 1996 that domestic demand will catch up with production capacity.[11] The presence of this mill is the primary reason why other companies have concentrated on the production of woodchip for export. Among the other major companies involved in the industry are Siampanmai and V.P. Eucalyptus Chip-wood. The latter is part of the Sahaviriya Group, which is also involved in the distribution of computers and office automation machines in Thailand, and which has strong connections with Japan and, especially, with Taiwan. In early 1990, the Sahaviriya Group announced that it planned to invest 10,000 million baht in eucalyptus plantations in Thailand, covering an area of 300,000 rai, through V.P. Eucalyptus Chip-wood.[12] The plan was soon stalled, however, as a result of scandals elsewhere in the industry.

Shell, whose head office in England, has eucalyptus plantations in Brazil and South Africa, and became interested in Thailand in the late 1980s. In September 1987, Shell requested promotional privileges from the BOI. The company proposed to invest 1,200 million baht on eucalyptus plantations occupying 125,000 rai in Chanthaburi.[13] Surveys of the initial area, however, revealed that only 30,000 rai was available within the RFD guidelines. The RFD then suggested that Shell survey 157,000 rai located elsewhere in the province.[14] Strong opposition to the project within Chantaburi (see discussion below), however, resulted in it being abandoned.

The development of eucalyptus plantations by such corporate interests have been subject to various scandals. The first of these involved a Thai-Japanese joint venture formed in 1987. In May 1987, the RFD approved a project called "The Joint Project of Reforestation between Public and Private Sectors in Honor of the King's Sixtieth Anniversary." The project was presented as being in honor of the King's sixtieth birthday in order to preempt opposition. A Thai-Japanese joint venture, Forest and Plantation Wood Industry, was formed in July 1987, prior to the invitation for bids to provide young eucalyptus plants by the RFD in August under the reforestation project. The company won contracts for ten million trees out of a total of 10.3 million.

The joint venture and its award of the contract soon became the subject of scandal as a result of conflict of interest involving high ranking personnel in the RFD and elsewhere in the Ministry of Agriculture and Cooperatives (MAC), the ministry within which the RFD is located. The problem began when the director general of the RFD, Chamni Boonyopas, threatened to abandon the project, claiming that the company had failed to deliver the trees as contracted. Pongse Sono, the deputy undersecretary of the MAC, and a former director general of the RFD with close links to the general manager of the joint venture, protested against the move to abandon the project. On 31 March 1988, a group of people claiming to be farmers, in conjunction with the Association to Promote Reforestation of the Private Sector, handed a petition to the MAC minister, General Harn Leenanond, with three demands. One of the demands was to sack the director general of the RFD within twenty-four hours.[15] While the evidence is rather sketchy, it appears that high ranking officials and business interests were involved in attempts to keep the project from being abandoned.

The above examples indicate considerable efforts by domestic and transnational firms to exert pressure on the Thai government—efforts which have received the cooperation of some members of the bureaucracy. Opposition, however, has come from environmentalists, villagers, and the public at large. Without such opposition, corporate interests would have gotten away with a great deal more than they have.

All Forms of Benefits to Eucalyptus Plantations

It is not difficult to understand why domestic and foreign private sector investors are interested in eucalyptus in Thailand. One of the most important reasons is that eucalyptus plantations receive generous incentives from the BOI. The incentives include duty exemptions on imported machinery and raw materials and various tax holidays and tax exemptions for extended periods. These incentives put eucalyptus at an advantage compared to other agricultural crops (including perennials), which receive no such promotion.[16]

The plantations also receive incentives in the form of very low rates for their leases. The original rate was one baht per rai per year! The rate was raised to a still very meager ten baht per rai per year following public protests against the plantations. This rate contrasts with the normal market rental rate to individuals and companies for eucalyptus plantations on reserved forests land of 150-200 baht per rai per year.

Even where companies must buy land from squatters to establish eucalyptus plantations, there is, in effect, a subsidy, since land with no title is bought at one-half to one-third of its market price. The term "squatters" for those who occupy forest land with no title is somewhat misleading. When the Reserved Forest Act was passed in 1964 more than eighty percent of the land in Thailand was without title. In establishing the forest reserves, many of those who lived on the lands, and who were "owners" according to traditional rights, were turned into squatters on what they considered to be their own lands.[17] Thus, when the RFD gives concessions for eucalyptus plantations to corporate interests, it is often on land already occupied by traditional owners. When such people face eviction or are coerced to accept compensation at below market rate, they frequently resist.

Unlike the so-called squatters, whose land could be taken away at any time by the government, those with eucalyptus plantation concessions were given guarantees by the RFD that their concessions will be secure. The situation for squatters began to change in 1988, when the government announced steps to provide titles to those already settled on their lands.[18] There is also the expectation, however, that eucalyptus planters will also eventually receive ownership title to their leased or purchased land.

Whether they receive ownership titles to the land or not, those with concessions for eucalyptus plantations are able to make substantial profits in a number of ways in addition to producing eucalyptus on land leased at cheap rates. To begin with, they are able to make money by clear-cutting the remaining forest on the land before growing eucalyptus. In fact, unofficially, officers of the RFD frequently permit concessionaires to have access to relatively fertile areas of reserved forest pending completion of negotiations. For example, in 1990, workers of Suankitti Co., Ltd. were arrested and charged with encroaching on reserved forest land and illegally logging forty valuable trees on the property.[19] Concessionaires are also able to convert part of the area under concession for use as tourist resorts, country clubs, and golf courses.

The Environmental Impact of Eucalyptus

Opposition to eucalyptus plantations by small farmers is associated not only with encroachment on their lands by the plantations, but also with direct observation of the detrimental environmental impact of eucalyptus. A study by the Thailand Development Research Institution (TDRI) in 1990 of farmers and small-scale eucalyptus planters revealed that many had complaints about the negative environmental impact of eucalyptus, including damage to crops and a reduction of soil moisture and water supply.[20] Most of the farmers in the study wanted the government to promote tree species other than eucalyptus in its reforestation projects.

Prior to the TDRI study, eucalyptus already had been subjected to criticism on environmental grounds from a variety of sources in Thailand. Rachanee Songanok has argued that the most undesirable feature of big eucalyptus plantations is their impact on the underground water table since eucalyptus is capable of absorbing large quantities of water at a much deeper level than other trees.[21] A similar point is made by Chakrit Homjun, who found in northeastern Thailand that underground water tables inside and outside the plantation differed by at least ninety centimeters.[22] The water absorption properties of eucalyptus make it unsuitable for reforestation, especially in watershed areas where trees are needed to help retain moisture. This property also makes it less appealing than other trees for providing shade or moisture for cash crops.

Experimental findings in the latter study also indicated that eucalyptus leaves mixed in the soil reduced the fertility of many seeds nursed in the soil.[23] When the leaf content was relatively high the growth of many plants was significantly reduced.

The Ministry of Agriculture and Co-operatives (MAC) issued guidelines for suitable land on which to grow eucalyptus on March 1988. The guidelines indicated that eucalyptus should be planted where the soil had a low potential for other crops. This included land which was arid, with rainfall of less than sixty days a year.[24] It is, therefore, important to note that much of the land identified as suitable for fast-growing tree plantations in a 1989 study by TDRI does not meet the MAC guidelines.[25]

Another concern about eucalyptus is that it is much more susceptible to forest fires than tropical forest trees. The use of eucalyptus on a large scale, therefore, greatly increases the danger of major forest fires. The danger is particularly great since people living near the trees may not be aware of the danger and traditional practices of slash-and-burn agriculture could result in forest fires.

The King himself expressed reservations about planting eucalyptus in 1987, directly to the media and through then prime minister Prem.[26] He noted his concern that the rapid propagation of eucalyptus could have adverse effects on the soil, on plants already existing in the area, and on wildlife.

Despite such drawbacks, in his concluding remarks at a 1987 government-sponsored conference on "Whether Eucalyptus is a Dangerous Tree?," the minister of the MAC, General Harn Leenanond, stated: "Eucalyptus is not a dangerous tree. Its negative side effect is not yet in evidence. Growing eucalyptus in a big plantation requires further studies."[27] Following the conference, the government launched a promotion campaign for eucalyptus plantations on a grand scale.

Protests Against Eucalyptus

After the above seminar, opposition to eucalyptus in the print media subsided. However, numerous protests took place in villages around Thailand. Protests had begun a short time before, in 1985, when 2,000 people of Seo sub-district, Uthumpornpisai district, Buriram province, moved into a eucalyptus plantation, pulling up young trees and burning eucalyptus nurseries and a government office. The protesters demanded withdrawal of a the concession which had been given to private planters without following proper procedures. Then, in December of the same year, people from Nongboa and Kokemanuang subdistricts, Prakam district, Buriram, marched against a government eucalyptus reforestation project which had resulted in their being evicted from their lands. They demanded the land back and agreed to grow eucalyptus or rubber trees if required by the government.

But it was in 1987 that protests broke out like forest fires. Generally, the protests resulted from encroachment into agricultural land or from a desire to protect existing forests from being cut to make way for eucalyptus. In February 1987, people from Noansawan sub-district, Prathumrat district, Roi Et province, together with people from seven or eight villages in the vicinity, pulled up young eucalyptus trees and seized a government tractor. They demanded that the natural forest, which was about to be clear cut for a eucalyptus plantation by the government, be kept in its original state so that they could use its products to help meet their daily needs. In April, people from Khampia sub-district, Trakarnpeuchpol district, Ubon Ratchathani province, demanded the complete stop of eucalyptus reforestation in the area since it was encroaching on their agricultural lands. In May, people of Yangkam sub-district, Ponsai district, Roi Et province, demanded a stop to the clear cutting of Dongbang forest for eucalyptus reforestation since it was the only forest left for the community. Before the end of 1987, three more cases were reported to have taken place in Yasothorn and Surin provinces.[28]

Protests continued in 1988. The fiercest fight took place when a protest was staged in Prakan district, Buriram province, in March after four people were arrested for growing fruit trees in an area in Dongyai forest reserved for eucalyptus. Some 2,000 people burned eucalyptus grown by the RFD in an area of twenty rai and also burned eucalyptus nurseries. The protest developed into a direct armed confrontation with the government when the people, feeling that they had nowhere else to retreat, decided to fight.[29]

Opposition to Shell's proposal to grow eucalyptus in Chantaburi (discussed above), began when a group calling itself the Study Group of Agricultural Problems wrote a newspaper article in late 1987 questioning the project.[30] One of the questions raised was whether some of the land with rainfall averaging between 1,800 and 2,500 millimeters could not be put to better use. Shell responded in the same newspaper in January 1988, stating that to be viable the plantation required that the eucalyptus be planted in an area with relatively good soil and rainfall.[31] It also pointed to other advantages of the location, such as its proximity to a planned deep-sea port. Then, in May, a member of parliament from Chantaburi belonging to the Thai People Party organized a protest march against Shell.[32] Most of those involved in the march were fruit and rubber tree growers. Finally, Shell decided not to go ahead with the project.

While protests against eucalyptus plantations have subsided since 1988, opposition among small farmers, environmentalists, and others continues.

Conclusion

Proponents of eucalyptus have paid little attention to studies pointing to the negative consequences of widespread planting of the trees. For many, the lure of profit has made them ignore broader social and developmental concerns. Those in government have been blinded by their association with those in the industry and by their inability to see the wisdom of local communal efforts to solve problems of deforestation.

Eucalyptus was initially introduced for reforestation. But it also came to serve as a cash crop for the pulp industry. The RFD has tried to combine the two objectives, while industry has only been concerned with the second. Are the two goals obtainable? To answer this, first we must determine whether eucalyptus can be of use for reforestation. In many settings, especially in watershed areas, the answer clearly is no. However, eucalyptus can be useful when grown in arid areas and in others such as in the northeast where underground water is salty and eucalyptus can help to improve water quality. The problem, from the perspective of the pulp industry, is that it is not profitable to grow eucalyptus in such areas.

Is it possible, then, to develop a reforestation program without the help of private business? Provided that the bureaucrats in the RFD can change their attitudes toward people in rural communities, the answer is yes. There is considerable evidence that people in these communities, if given proper authority and responsibility, can look after the natural forests around them in a way that also provides substantial benefits to the community. These communities can support themselves much better from the natural forest than surrounded by eucalyptus planations. Community-based forestry provides a sustainable alternative for reforestation to reliance on eucalyptus plantations.

Finally, what of the need to provide raw materials to meet the rapidly increasing requirements of the pulp industry? This should be left to the market to work out its own solution without the incentives and interference that has characterized the promotion of eucalyptus planations to date. The government should pay more attention to finding long term solutions to environmental problems such as deforestation. It should not interfere with market mechanisms, especially with the main aim of protecting certain business interests. Unfortunately, the persistence of strong patron-client and other close relations between those in government and those in the private sector makes this difficult.

Notes

1. Dow Mongkolsmai, Wachareeya Tosanguan, and Chayan Tantiwasdakarn. *Pulp and Paper Industries in Thailand*. Report prepared by Pacmar, Inc., in association with A&R Consultants Co., Ltd., 1987.

2. The discussion of pulp industries in this chapter is limited only to short fibre pulp since Thailand cannot produce long fibre pulp and industrial paper uses only short fibre pulp.

3. Anat Arbhabhirama, *et al. Thailand Natural Resources Profile* (Singapore: Oxford University Press, 1988).

4. Royal Forestry Department. *Annual Report of 1988* (Bangkok: Ministry of Agriculture and Co-operatives, Royal Forestry Department, 1988).

5. Arbhabhirama, *Thailand Natural Resources Profile*.

6. *Siam Rath*, 10 August 1987.

7. *Matichon*, 26 June 1988.

8. *Khao Parnich*, 29 September 1987.

9. *Siam Rath*, 10 August 1987.

10. *Sethkij*, 4-10 February 1989.

11. *Matichon*, 7 May 1990.

12. *Sarnsontes*, January 1990.

13. *Matichon*, 26 November 1987.

14. *Prachachart Dhurakij*, 30 May 1988.

15. *Matichon*, 26 June 1988.

16. Thailand Development Research Institution. *Quarterly Review*, Vol. 5, No. 2, 1990.

17. See Anan Ganjanaphan, "Community forestry management in northern Thailand," in Amara Pongsapich, Michael C. Howard, and Jacques Amyot (eds.), *Regional Development and Change in Southeast Asia in the 1990s* (Bangkok: Chulalongkorn University Social Research Institute, 1992), pp. 75-84.

18. Communities were divided into three categories: those that existed before 1967, those that came into existence between 1967 and 1975, and those coming into existence from 1975 to 1981. The first group will receive proper titles, the second group will receive deeds allowing them to use the land and to pass it on to their heirs (but not to sell it to outsiders), while the third group will be allowed to use the land temporarily (with the expectation that their title will be upgraded over time).

19. *Siam Jodemaihet*, 27 April-3 May 1990.

20. Thailand Development Research Institution. *Quarterly Review*.

21. Rachanee Songanok. "Commercial eucalyptus plantation in the province of Cha-Choengsao," in *Journal of Agricultural Economic Research* (Office of Agricultural Economics, Ministry of Agriculture and Co-operatives, Bangkok), Vol. 12, No. 36, 1990.

22. Chakrit Homjun, *et al. Impact of Eucalyptus Plantation on Soil Properties and Subsequent Cropping on Northeast Thailand*. Research paper prepared with the support of USAID and RDI of Khon Kaen University, 1989.

23. Homjun. *Impact of Eucalyptus Plantation*.

24. Songanok. "Commercial eucalyptus plantation."

25. Thailand Development Research Institution. "Area for fast-growing tree: Whether it is available—an example of using geographic information system to locate the suitable area for fast-growing tree plantation." Paper No. 5 presented to a seminar on "Economic Forest: Reality or Dream," Bangkok, 23 February, 1989.

26. *Matichon,* 1 March 1987, and 1 April 1987.

27. *Khao Parnich,* 29 September 1987.

28. *Matichon,* 19 May 1988.

29. *Matichon,* 26 March 1988.

30. *Matichon,* 26 November 1987.

31. *Matichon,* 13 January 1988.

32. *Prachachart Dhurakij,* 30 May 1988.

8

Biotechnologies and Sustainable Development: Potential and Constraints

Nagesh Kumar

It is now widely recognized that many of our current crises such as high levels of environmental pollution, global warming, depletion of the ozone layer, desertification, depletion of non-renewable resources, loss of valuable biological diversity, soil degradation, concentration of wealth and income, and rising unemployment and poverty have something to do with the nature of technologies developed and employed in industry and agriculture. The Green Revolution technology in agriculture involving the use of high yielding varieties of seeds, high doses of chemical fertilizers, pesticides and other chemical inputs, irrigation, and mechanized equipment resulted in a massive growth of agricultural produce. But these technologies provided a poor basis of long term food security. Because of the high dependence on purchased inputs, these technologies favored only big farmers and resulted in a widening of regional and intra-regional disparities in developing countries. This new technology is heavily dependent upon non-renewable resources (oil based chemicals, fertilizers, and energy). The large scale application of such inputs has led to an erosion of the long term productive potential of land and to a loss of genetic diversity that is crucial for sustained growth of agricultural output. Similarly, technological changes in industry have tended to be capital deepening and intensive in the consumption of non-renewable resources. Moreover, related economies of scale and scope have favored concentration. It is within this context that we must assess the emergence of biotechnologies which have the potential of contributing to a more sustainable pattern of development.

This chapter outlines the potential of biotechnologies in providing a basis for sustainable development especially in the context of developing countries. It then examines the constraints emanating from the emerging trends in biotechnology research and development for the exploitation of this potential. It concludes by outlining some policy imperatives.[1]

Potential of Biotechnologies in Fostering Sustainable Development

Food Security

By the turn of the century the world population will have risen to over six billion. To merely maintain the current consumption levels, an increase of twenty-six percent in the world's average grains yields will be needed. The bulk of this increase will be required in developing countries. Unlike in the past, when yield increases were achieved under favorable cropping conditions by the Green Revolution, future improvements in yields must come from the productivity improvement of traditional farmers cultivating low-yield crops under marginal conditions and with little dependence on purchased inputs. Integrated with conventional plant breeding, biotechnologies can play an important role in providing the basis for economic and ecological sustainability of further increases in agricultural yield and thus ensure food security for developing countries. Some illustrations in this context are as follows:

(1) Biotechnology can help improve agricultural productivity by raising the ceiling of yields through enhancement of the efficiency of photosynthesis.[2] It can also help in bridging the gap between actual and potential yields by accelerating the pace of plant breeding. Tissue culture techniques, for instance, can be used for rapid multiplication of single superior and virus-free and disease-free (elite) plants and the regeneration of plants difficult to propagate sexually. Phenomenal gains in yield have already been reported in different parts of the world through the application of tissue cultures in oil palm, coconut, banana, tubers, and other plants.[3] Techniques like embryo rescue, protoplast fusion, and the use of DNA vectors enable plant breeders to overcome the barriers of sexual incompatibility in transferring desirable traits.[4]

(2) Biotechnology can assume a crucial role in reclamation of poor soils and waste lands. For instance, in South Asia and Southeast Asia alone about 86.5 million hectares of land can be made productive with rice varieties tolerant to salinity and alkalinity.[5] In India, for instance,

biotechnology is being used to develop mustard plants tolerant to salinity as a part of a technology mission aiming to achieve self-sufficiency in edible oils. Similarly, adaptation of cereal crops to drought-prone areas could be valuable for solving the food problem in sub-Saharan Africa.

(3) Biotechnology can accelerate the development of low input agriculture. This will be of special significance for developing countries with large populations of small and marginal farmers. Low-input agriculture requires disease-resistant and pest-resistant varieties with an ability to fix nitrogen. Biological nitrogen fixation through various agents, such as blue-green algae that are symbiotic with the water fern Azolia and rhizobia in symbiosis with legumes, offers a great potential for increasing the agricultural yield in developing countries without increasing their dependence on chemical fertilizers.

(4) Biotechnology can also be used to redress the imbalances caused by the Green Revolution. The Green Revolution covered only a small band of cereals and did not benefit a wide range of crops produced and consumed by the world's poor—such as cassava, yams, and sweet potatoes, which are vegetatively propagated.[6] Biotechnology not only has a tremendous potential for improving the productivity of these crops but can also be applied to raise their nutritive value.

(5) Biotechnological techniques provide the basis for ecological sustainability of the high yields by reducing risks from pests, pathogens, and weeds and promoting conservation of genetic resources. The genetic material in wild species contributes billions of dollars yearly to the world economy through high yields and by saving crops from diseases, and through new drugs and medicines. These are crucial resources for ensuring the sustainability of the long term food security. The new techniques allow preservation of cloned DNA and materials having DNA in their native state for genetic conservation. Further, somaclonal variation presents opportunities for adding to the pool of genetic variability.

(6) The potential spectrum of biological pesticides naturally occurring or in the form of genetically restructured organisms that are specifically pathogenic to important pests, parasites, or weeds is enormous and remains largely unexplored.[7] Integrated pest management (IPM) which reduces the need for agrochemicals through the use of natural pesticides and integrated cropping patterns is especially appropriate for small and marginal farmers in developing countries.

Resource Saving in Industry

A great variety of food processing and chemical processes such as fermentation involve the use of microbes. The efficiency of these processes and the nutritive value of the product can be increased through selection of more productive microbial strains, control of culture conditions, and through adaptation of the fermentation processes. These techniques can be employed not only for treatment of affluents, but also for production of valuable products from industrial wastes. Thus, these technologies can help in solving pollution problems while also generating additional value added. For instance, single cell proteins can be produced from molasses, paper mill affluents, or hydrocarbons and can be used as high protein cattle feeds.[8]

More Efficient Exploitation of Mineral Resources

Metals can be extracted from low grade ores with biotechnological techniques. Thus, modified bacteria can be used to accelerate production of chemical solutions that wash out normally insoluble mineral compounds containing such metals as copper, zinc, nickel, and lead. Bacterial leaching can be used to extract the remaining minerals from mining and metallurgical wastes. It can also help in extracting the remaining oil from the dead oil wells.

Fostering Decentralized Rural Industrialization

With biotechnological advances, it is possible to integrate agriculture with the production of food, animal feeds, energy, fertilizer, and a number of industrial products. Such integrated systems can help to generate valuable resources from agricultural wastes and to foster decentralized rural industrialization in developing countries. The recycling of wastes into energy helps reduce the pressure on conventional sources of energy. Certain encouraging experiences from Brazil, China, and India with these food-energy integrated systems are now available and could be extended to other developing countries.[9]

Saving of Foreign Exchange Resources

Most of the developing countries suffer from a perpetual balance of payments crisis. Another aspect of sustainability for them is the potential of saving foreign exchange resources. Through the development of crops requiring lower inputs of chemical fertilizers, pesticides, and herbicides, by contributing to the self-sufficiency of food, and by generating energy and chemicals from the recycling of biomass and other wastes, bio-technologies can help developing countries to save scarce foreign exchange resources. Therefore, they can contribute to a more self-reliant and, hence, more sustainable pattern of development in the developing countries.

Population and Health

Population explosion is a major developmental problem in most developing countries. Biotechnologies can be of great help in the efforts of these countries to reduce the birth rate. Biotechnological techniques are being employed to develop better and more effective kits for early detection of pregnancy and infertility vaccines in India.

In the areas of health and medicine, biotechnologies are leading to the development of new drugs (e.g., TPA—a vital life-saving medicine for heart attack) and to production of more efficient substitutes of known drugs (e.g., humulin—human insulin). These techniques also have a tremendous potential in immuno-diagnostics and the development of vaccines.

Reforestation and Control of Pollution

Some of the biotechnological techniques, for instance tissue culture, are invaluable for programs of reforestation and social forestry. They enable the rapid regeneration of tree saplings and the faster growth of trees. They also can improve fiber quality, disease resistance, and growth under adverse physical conditions. Further, these techniques provide more efficient means of fighting the menace of urban and industrial pollution. With them, detoxification of almost any substance is conceivable. One of the first modified microorganisms to be patented was actually a bacteria capable of eating oil spills.

Constraints

Biotechnologies thus appear to have a tremendous potential for fostering a more sustainable pattern of development. Given a proper policy framework for their development and exploitation, they can provide developing countries with a valuable means to expedite the process of their development and to pursue a pattern of growth with positive income distribution effects. However, the potential of biotechnology in evolving a sustainable pattern of development is likely to remain unutilized in view of the emerging character of the biotechnology industry in the industrialized countries.

Emerging Biotechnology Industry

Almost all of the biotechnology research and development activity is concentrated in the industrialized countries—the United States, Japan, and Western Europe. The emerging biotechnology industry in these countries is characterized by the following significant trends which have implications for developing countries.

(1) Privatization of Research. One of the most noticeable features of research and development in biotechnology is its high degree of privatization. Unlike in the case of the Green Revolution, little biotechnology research is in the public domain. Research in biotechnology was spearheaded by small specialist venture-capital firms floated by professional scientists. Gradually, transnational corporations, especially those with primary interests in pharmaceutical, chemical, food, and energy industries, became interested in biotechnology. Besides setting up in-house research and development programs, these transnational corporations have established strong links with specialist biotechnology firms through takeovers, joint ventures, or research contracting. Transnational corporations have also contracted much of the biotechnology research carried out in universities and publicly-funded research institutions and, thus, enjoy the right of prior access and patent ownership.[10] In the United States, for instance, nearly half of the companies engaged in biotechnology research have arrangements with universities. One-fourth of all biotechnology research at United States universities in the mid-1980s was supported by the industry.[11] An immediate consequence of this trend is that there is little flow of basic scientific information and knowledge through the usual scientific channels, such as publication in learned journals, conferences, and other professional interactions.

(2) Vertical Integration. The secrecy surrounding biotechnology research and the high dependence of agrochemical sales on the future course of biotechnology research has led to the trend of vertical integration among transnational corporations, especially agrochemical conglomerates taking over specialist biotechnology companies on the one hand and seed companies on the other. Agrochemical transnational corporations such as Sandoz (Switzerland), ICI (U.K.), Rhone-Poulenc (France), Upjohn (U.S.A.), Dekalb-Pfizer (U.S.A.), and Ciba-Geigy (Switzerland) now rank among the top ten in the seeds business. Such vertical integration is aimed at consolidating a monopoly position through use of the marketing infrastructure of seed companies for the sale of genetically modified seeds which respond only to particular brands of herbicides or pesticides.

(3) Proprietary Ownership. Another marked trend is that of increasing proprietary ownership of biotechnology-based processes and products through the use of industrial patent laws and plant breeder's rights to maintain a monopoly position. Though, historically, living organisms have not been subject to patent protection, there has been an increasing tendency in the industrialized countries in recent years to provide patent protection to the biotechnology-based innovations through modifications of laws. In 1980, the United States Supreme Court allowed patenting of a modified microorganism. In 1986, a plant variety was granted a utility patent like an industrial invention in the United States (so far only plant breeders' rights were available to new varieties). In 1987, the United States Patent Office announced the industrial patenting of higher life forms, including pets and livestock.[12] The industrialized countries are seeking universal recognition of these intellectual property rights through multilateral trade negotiations (TRIPS in the new round of GATT negotiations) and by unilateral pressures on developing countries (such as through Special 301).

Implications for Developing Countries

The increasingly private and proprietary character of biotechnology research has important implications for the access of developing countries to these technologies and the determination of research priorities.

(1) Access of Developing Countries. The growing commercial character of university and publicly-funded research institutions, together with trade secrecy, suggest that the traditional means of transfer of technical information, such as interaction among scientists and publications in learned journals, will become insignificant. The closely held nature of the technology would make for highly imperfect markets and, therefore, the prospects of its transfer on reasonable terms and through appropriate channels appear remote.[13]

(2) Research Priorities. Research priorities in biotechnology are likely to be determined by commercial prospects and the global strategies of transnational corporations rather than by what is desirable for the poor in developing countries. As observed above, biotechnologies can be fruitfully employed to develop varieties which require lower inputs of agrochemicals such as fertilizers, herbicides, and pesticides that would be more appropriate for millions of small and marginal farmers in the developing world. Instead of developing such varieties, the agrochemical transnational corporations such as American Cynamid, Monsanto, Ciba-Geigy, Rhone-

Poulenc, Du Pont, and ICI, which have taken over most of the seed companies, are contracting research to develop seeds resistant to their proprietary herbicides, hence making them more dependent on chemicals. Monsanto, for instance, is engaged in the development of seeds that are resistant to its patented herbicide "Round Up." One such soybean variety alone is likely to increase the sale of Round Up by about US$150 million a year. Research is underway to incorporate such resistance in maize and sorghum.[14]

(3) Substitution of Commodity Exports. A significant part of the biotechnology research in Western countries is actually aimed at industrial production of certain high-value plant secondary metabolites such as flavors, fragrances, and medicinal plants and at other commodities like sugar and gum arabic, presently imported from developing countries. These developments, like those resulting from other frontier technologies such as flexible automation systems and the creation of new materials, will further diminish the place of developing countries in the international division of labor. It has been estimated that the developing world may lose about US$10 billion of its annual export earnings from biotechnology-based commodity substitutions in the near future.[15] The loss of export earnings would only aggravate the already precarious condition of most developing countries. It could prove to be particularly devastating for a number of African and Caribbean economies having high dependence upon exports of these commodities.[16]

(4) Local Technological Capability and Plant Breeding. The trend of granting intellectual property rights and seeking their universal recognition on all biotechnology based livestock and plant breeding has important implications for developing countries. Once accepted, these intellectual property rights would imply the granting of an exclusive monopoly over the import, manufacturing, and selling of any plant varieties or livestock containing patented traits to companies owning such rights. Even the storing of seeds by farmers for further use out of their output, which is a widespread practice in the developing world, will contravene these monopoly rights. Patenting in this area has been found particularly objectionable by developing countries because the basic raw material for genetic manipulation are the wild genetic resources which are heavily concentrated in tropical or subtropical areas. These vital resources have been appropriated by the industrialized countries without any compensation.[17] Therefore, granting a monopoly right to a company on a variety for specific traits is tantamount to the use of the developing world's resources for commercial exploitation by transnational corporations worldwide without the source countries ever sharing any part of the

quasi-rent. Besides, in this area, intellectual property protection would deny to developing world countries whatever little access and chance they have to develop an indigenous biotechnology capability and industry to serve their priorities and needs. This would, in particular, cast a fatal blow to the plant breeding activity in the developing world.[18]

(5) Socio-Economic Imbalances. Developments in biotechnology will lead to concentration on a few materials. For instance, quadrupling of palm oil production will see a proportionate rise in its share in global edible oil consumption at the cost of others such as sunflower, cocoa, soya, and cotton seed. Such developments will have wide-ranging socio-economic repercussions. For example, the introduction of herbicide-linked seeds currently being developed by transnational corporations in developing countries will cause the displacement of predominantly female labor employed in the weeding activity.[19]

The emerging character of the biotechnology industry, therefore, does not allow its utilization for promoting sustainable development. Instead, the research priorities distorted to serve the commercial interests of transnational corporations that dominate it are aiming to increase the dependence of agriculture in developing countries on chemical inputs and to provide developed countries with substitutes for the commodity exports of developing countries. Further, the increasingly private and proprietary character of biotechnology research will serve to widen the technological gap between the rich and poor nations. Technology is now being used by countries as a new means of economic domination.

Countries in the developing world, therefore, will have to strive on their own if they are to be able to exploit the positive potential of biotechnology. They will have to build up biotechnology research geared to their needs and priorities and utilize it for their benefit. Some initiatives have already been undertaken in this direction in the framework of the United Nations system and at national levels by a few developing countries. International action is urgently called for to help the developing countries to absorb the loss of export earnings as a result of developments in biotechnology.[20]

Concluding Remarks

In the context of recent concern about the sustainability of the current pattern of economic development, biotechnologies present a ray of hope. Given the proper focus, orientation, and organization for their development and utilization, they can prove to be valuable in fostering a more sustainable pattern of development. However, the character of biotechnology research and industry that is emerging in the industrialized countries suggests that rather than exploiting its potential in promoting equitable and sustainable development, it will be used to serve the commercial interests of transnational corporations and as a new means of technological domination of the developing world.

International cooperation, therefore, is of immediate importance to reverse the emerging trends of privatization and proprietary ownership of the new knowledge to ensure its utilization for maximum common advantage of the world economy. Developing countries have a special challenge to build up local technological capability in this area to exploit the tremendous potential that biotechnologies offer to them. In their effort to build up local capability, mutual cooperation among developing countries has been shown to have an important role to play.[21] In addition, their solidarity in international negotiations, hopefully, can stall some of the adverse trends and initiate international intervention to gear biotechnology research into more desirable priorities.

Notes

1. An extensive annotated bibliography on the issues covered in this paper has been published separately by the Research and Information System for Non-aligned and Other Developing Countries, *Biotechnology Revolution and the Third World: An Annotated Bibliography* (New Delhi: Research and Information System for Non-aligned and Other Developing Countries, 1988).

2. M.S. Swaminathan, "Biotechnology and sustainable agriculture," in Research and Information System for Non-aligned and Other Developing Countries, *Biotechnology Revolution and the Third World: Challenges and Policy Options* (New Delhi: Research and Information System for Non-aligned and Other Developing Countries, 1988), pp. 33-34.

3. Albert Sasson, "Promise in agriculture: Food and energy," in Research and Information System for Non-aligned and Other Developing Countries, *Biotechnology Revolution and the Third World*, pp. 55-88.

4. Swaminathan, "Biotechnology and sustainable agriculture."

5. M.S. Swaminathan, "Biotechnology research and third world agriculture," in *Science*, Vol. 218, December, 1982, pp. 967-972.

6. Iftikhar Ahmed, "Pro-poor potential," in Research and Information System for Non-aligned and Other Developing Countries, *Biotechnology Revolution and the Third World*, pp. 134-149.

7. Swaminathan, "Biotechnology and sustainable agriculture."

8. Sasson, "Promises in agriculture."

9. Ignacy Sachs, *Resources, Employment and Development Financing: Producing without Destroying—The Case of Brazil*, RIS Occasional Paper No. 30 (New Delhi:

Research and Information System for Non-aligned and Other Developing Countries, 1988), pp. 967-972.

10. David Dembo, Clarence Dias, and Ward Morehouse, "The vital nexus in biotechnology: The relationship between research and production and its implications for Latin America," mimeo (Caracas. Venezuela, 1989); Frederick H. Buttel and Martin Kenney, "Prospects and strategies for overcoming dependence," in Research and Information System for Non-aligned and Other Developing Countries, *Biotechnology Revolution and the Third World*, pp. 315-348.

11. Dembo, Dias, and Morehouse, "The vital nexus in biotechnology."

12. Mooney, Pat Roy. 1988. "Biotechnology and the North-South conflict," in Research and Information System for Non-aligned and Other Developing Countries, *Biotechnology Revolution and the Third World*, pp. 243-278.

13. Dembo, David and Ward Morehouse, *Trends in Biotechnology Development and Transfer*, IPCT. 32, Technology Trends Series No. 6 (Vienna: UNIDO, 1987); Gerd Junne, "Bottlenecks in the diffusion of biotechnology from the research system into developing countries' agriculture." Paper presented at the 4th European Congress on Biotechnology. Amsterdam, June, 1987.

14. *Biotechnology and Development Monitor*, Vol. 1, No. 1 (Amsterdam, 1989).

15. V.R. Panchamukhi and Nagesh Kumar, "Impact on commodity exports," in Research and Information System for Non-aligned and Other Developing Countries, *Biotechnology Revolution and the Third World*, pp. 207-224.

16. Gerd Junne, "Incidence of biotechnology advances on developing countries," in Research and Information System for Non-aligned and Other Developing Countries, *Biotechnology Revolution and the Third World*, pp. 193-206.

17. Jack Kloppenburg, Jr., and Daniel Lee Kleinman, "The genetic resources controversy," in Research and Information System for Non-aligned and Other Developing Countries, *Biotechnology Revolution and the Third World*, pp. 279-314.

18. South Commission, *Statement on the Uruguay Round* (Cocoyoc, Mexico: South Commission, 1988).

19. Ahmed, "Pro-poor potential."

20. Nagesh Kumar, "Biotechnology revolution and Third World: An overview," in Research and Information System for Non-aligned and Other Developing Countries, *Biotechnology Revolution and the Third World*, pp. 1-30.

21. Kumar, "Biotechnology revolution and Third World."

9

Energy Strategies and Environmental Constraints in China's Modernization

Mark McDowell

The conflict—or, in more traditional Maoist terms, the "contradiction"—between China's growing energy needs and the health of its environment has become increasingly acute in recent years.[1] In a nation where energy shortages already impose bitter hardships on hundreds of millions of peasants and cause substantial industrial capacity to lie idle, China's official goal of quadrupling GNP between 1980 and 2000 will demand sizeable and sustained increases in energy consumption. While it may be technically possible to achieve these gross increases, the development of the nation's energy resources along present lines will cause environmental degradation on a wide scale. Moreover, the consumption of such quantities of energy will inflict environmental costs on the Chinese people which may be both economically prohibitive and even socially destabilizing.

A dramatic illustration of these new complexities has been the controversy over the Three Gorges Dam Project (TGP) which calls for the building of a 175 meter high dam on the Changjiang (Yangtze) River and the installation of the world's largest hydroelectric generating project. The TGP proposal caused an outcry among the Chinese scientific community and attracted the international attention of environmental critics. Its huge scale set in motion a politically fascinating series of debates which spilled over from strictly scientific questions to criticism of government budgetary and development priorities, and an unprecedented amount of open special interest group lobbying. Environmental criticisms such as the potential impact of the project on downstream water quality and quantity, river morphology, flooding of upstream land, forced relocation of approximately half a million peasants, and numerous others have for perhaps the first time played a significant role, alongside more customary budgetary or political criteria, in postponing the project. The project has been shelved numerous times, most recently in January 1989, yet the TGP has, in the best tradition of Chinese politics, made repeated comebacks from political disgrace—being approved yet again at the Seventh National People's Congress in March 1992.

Some Western critics may attribute this stubbornness in the face of overwhelming environmental evidence to pharaohic tendencies in the Chinese leadership or to the residual "ferrophilia" of Stalinist development ideology. Similarly, China's rapidly expanding coal use has been branded an irresponsible environmental hazard on a global as well as national scale since Chinese greenhouse gas emissions account for about ten percent of the global total, and are growing proportionally. Environmentalists, however, have generally been willfully blind to the better intentions of the Chinese leadership—to bring economic development and a higher standard of living to the Chinese people—and have ducked the issue which these energy strategies are intended to solve. That is, given that China needs energy to develop, what are China's alternatives, and how do comparable plans of energy development measure up to the TGP in terms of environmental hazard and degradation. This chapter will outline in qualitative terms the environmental consequences of energy strategies available to China, and indicate strategies which might mitigate the environmental degradation accompanying energy development.

Environmental Policy in China

Before examining energy and environmental policy alternatives, we ought to be aware of the political climate in which policy-making takes place.

After discussions of the Chinese environment, ranging from the critical to the scathing, which have appeared in recent years,[2] the image of the naive 1970s China watcher credulously relaying government propaganda about regreening barren slopes and recycling human excrement, has itself become something of a caricature. The naive faith of Western observers in Chinese environmental virtue has, like many of the other political longings which were projected onto Mao's China, become a casualty of China's open door policy. The Chinese themselves now admit that gross mismanagement of the environment has occurred over the past three decades and that these mistakes are now costing China dearly.[3] Yet, a recognition of past errors and a wholesale abandonment of Maoist policy alone will not bring sudden environmental salvation anymore than it can completely solve problems of industrial production or bureaucratic inertia.

Centrally planned economies might be expected to be more able to curb the damaging effects of competition for resources and externalization of environmental costs, and to ensure a strict supervision of industry and an environmentally optimal degree of consumption. This has not, however, been borne out in the historical experience of either Eastern Europe or China. Rather, the environmentally salient characteristics of Marxist states has been a "plan mentality," a focus on quantitative output indicators above all other considerations,[4] combined with an endemic underpricing of industrial inputs, and a lack of open popular dissent which can serve to warn of serious environmental costs. The Chinese experience has been particularly unfortunate as it has often exhibited traditionally capitalist faults such as quick profit-taking during some of its more market-oriented historical phases. The present phase of economic liberalization since the end of the 1970s has seen the government of a transitional economy grappling with the characteristic problems of capitalism and communism simultaneously.

The Chinese are, similarly, grappling with a number of different approaches to environmental regulation, shedding Mao's campaign-oriented mass mobilization strategies, attempting to institutionalize bureaucratic regulation, and trying to graft market regulation principles onto an economy to which they are ill-suited.

Energy and the Environment

China's most severe environmental problems arise directly or indirectly from energy procurement and use. Energy (or lack of energy) causes environmental degradation, and environmental degradation can, in turn, exacerbate energy shortages. Chinese planners identified energy along with transportation capacity as two critical bottlenecks to economic development very early in the reform period, but environmental degradation may, in the long term, jeopardize any gains of development and act as an equally formidable limiting factor. Environmental considerations place very strict medium-term to long-term constraints on energy procurement, and it therefore should be a priority that Chinese leadership integrate energy and environmental planning in order to make the achievement of the "Four Modernizations" possible. In order to understand China's environmental dilemmas, therefore, we must first understand its rather unique energy problems.

Despite its poverty, China is a very energy rich nation. China ranks third behind the former Soviet Union and the United States in total energy production[5] and possesses the second largest coal reserves and the largest theoretical hydropower potential in the world. In sum, China possesses ten percent of the world's energy reserves.[6] Yet, per capita consumption of energy is extremely low, less than a tenth of the North American average: twenty percent of industrial capacity lies idle because of power shortages and, as of 1984, forty percent of peasant households had no electricity.[7]

Part of the problem is that regional distribution of resources is very inconvenient, the overwhelming share of coal reserves and production lie in the north and northwest and of oil in the north and northeast. A similar imbalance exists in hydropower potential, with three-fourths located in the southwestern areas of Tibet, Yunnan, Guizhou, and Sichuan. Thus, the population and industrial centers of east and south China are without large-scale local power supplies.[8] Power shortages are exacerbated by China's notorious transportation bottlenecks. Bringing fossil fuels 1,000 kilometers south to the Yangtze places a tremendous burden on the transportation infrastructure, coal already being the largest single item of rail cargo. To make things worse, China appears to be beginning a permanent shift in spatial distribution of energy sources to more distant western sources in Xinjiang, Gansu, and Inner Mongolia, as well as Shaanxi, as the older northeastern oil fields and eastern and southern coal fields wind down.[9] This shift will further exacerbate transport problems and make the development of energy sources near sites of consumption more urgent.

The structure of China's use of energy is unique and is a key to environmental problems. About five percent of China's primary commercial energy consumption is derived from hydropower, seventeen percent from oil, and seventy-six percent from coal.[10] This huge share for coal is out of line with developed nations. In the United States and the former Soviet Union, for example, coal accounts for about thirty percent of primary commercial energy. Such high coal use is also unique among developing countries, for no other developing country has coal reserves of this magnitude.

Commercial energy alone does not provide a complete picture for China's non-commercial energy sector. The use of non-commercial energy sources such as wood and grasses is as important to the Chinese peasant as in other developing countries. Once such sources are taken into account, China's energy sources are actually about one-half coal and one-third biomass, with oil, hydroelectricity, and gas making up the remainder.

While commercial and non-commercial sectors are not exactly coterminous with the urban and rural division, China's urban and rural halves have different environmental problems, arising largely from different types of energy use—coal and biomass respectively. China is at an unfortunate stage of development where it must deal with the environmental problems of a modern sector at a time when it is still struggling with the environmental problems of a traditional rural sector which have plagued the country for over a thousand years.[11]

China's Modern Energy Sector

Given that coal is the overwhelming energy source of China's industrial sector as well as of urban households, what are the environmental consequences?

In aggregate terms, China's sulphur dioxide emissions are eight time as high as Japan's and particulate emissions are as high as those in OECD countries.[12] Coal burning accounts directly for eighty percent of sulphur dioxide and sixty percent of particles.[13] Chinese fossil fuel burning emissions account for a substantial and growing percentage of the world total, but ignoring any contribution to global environmental disaster, the effects on China are severe in the form of extremely high rate of air pollution. Average seasonal levels of pollutants in cities such as Beijing, Lanzhou, Chongqing, and others often exceed the levels for total suspended particles of sulphur dioxide and carbon monoxide deemed tolerable for long term exposure by one-hundred percent.[14] City dwellers must cope with high rates of cancer and respiratory illness, the disposal of a billion tons of coal ash, and the localized appearance of acid rain.[15]

Despite the obvious source of these problems in China's over-dependence on coal, there appears to be little possibility of significantly altering China's energy structure before the end of the Ninth Five Year Plan in the year 2000. Depressed world oil prices have dampened enthusiasm for exploration in the South China Sea and cast doubt on the possibility for any major expansion of oil production in the near future. Natural gas, even according to the most optimistic predictions, can only supply three percent of China's energy by the year 2000.[16] Hydroelectric power shows the greatest possibility for development since only six percent of potential presently is tapped, but the long lead times for large projects limit their short-term impact on urban and industrial needs. The TGP, to cite an admittedly extreme example, has a projected construction period of eighteen years. In the interim, effort must be focused on improving the utilization of coal to reduce both the volume burned and the emissions per unit burned.

China's Modern Energy Sector: Environmental Reforms

Although China's energy consumption is very low on a per capita basis, in aggregate it greatly exceeds that of Japan. Because China's GNP is only one-sixth of Japan's, energy use per unit of output is almost ten times as high. In fact, China's energy use per unit of output is far above that of any developed nation, more than twice that of the United States, and higher than that of almost all developing countries.[17] Any reduction of this inefficiency would reduce appreciably both the amount of coal burned and the amount of pollutants released by a given quantity of burning. While China's energy use per dollar of GNP declined by nineteen percent from 1973 to 1983, this improvement was slower than, for example, Japan's saving over the same period from a far more efficient base.[18] Conversion efficiencies for major types of industrial and domestic coal use in China are typically half to two-thirds of comparable OECD figures: in fact, China's overall energy conversion efficiency is roughly equivalent to that of developed nations in the 1950s.[19] Similarly, emissions per unit of energy used are much higher in China. In the mid-1970s, China emitted sixteen times as much TSP per kilogram of coking coal as the best furnaces of Japan and West Germany.[20]

These comparisons are significant for they show that China's reliance on vintage technologies is a crucial environmental problem. Throughout the 1960s and early 1970s, China followed a policy of what has been called "second-best technology," referring to the use of ten to thirty year old processes to build an indigenous industrial base cheaply, and without foreign control.[21] This autonomy was achieved through the use of low cost energy and by taxing the carrying capacity of the environment. Since the oil shocks, energy efficiency in the industrial economies has increased rapidly, and pollution control technologies are a much lighter cost burden than in the 1950s.[22] One might protest that China is wholly unable to bear the burden of these higher capital new technologies, but Thomas Rawski has pointed out that China's "weakest industrial sectors are precisely those that have failed to advance in the direction of capital intensive world technologies."[23] Pollution control is such a sector, and the logic of China adopting high-tech solutions is even stronger. Some of the steps that can be taken to conserve energy and reduce pollution are quite rudimentary—washing more coal to remove impurities before burning, for example. Still, the need to adopt up-to-date techniques to manage the environment casts fundamental doubt on development strategies which advocate autonomously developed "appropriate" or "second-best" technologies. Kinzelbach laments that, while Westerners waste energy through over-consumption, the Chinese, who can least afford it, waste it before it ever reaches the consumer.[24]

There is, unfortunately, another attractive development paradigm which is discredited by the Chinese experience—the "small is beautiful" argument. Industrial scale is a severe handicap to both efficiency and emission control. Conversion of coal in thermal generation is a major source of waste as

China relies on small (fifty to 200 megawatts) plants with an efficiency three-fourths that of larger (300 to 600 megawatts) Western plants.[25] By adopting more modern methods, the savings on a hundred billion kilowatt hours of thermal power generation would be fifteen million tons of coal.[26] In addition to creating substantial diseconomies of scale in terms of power generation, small and dispersed enterprises make the installation of pollution control technology costly and administrative supervision difficult. The problem of China's decentralized energy use has historical roots in the policy of local industry self-sufficiency advocated during the Great Leap Forward at the end of the 1950s. While this autarchy was encouraged for reasons of ideology and military security, the pattern is maintained by the desire of local authorities, inculcated by experience in a socialist "shortage economy," to have assured access to energy sources.[27] With the rapid economic growth envisioned by China's economic planners, however, China is now presented with a unique window of opportunity to rectify past errors and reduce long-term economic costs of pollution by emphasizing pollution control technology in its new capital stock.

Some aspects of economic reform are working against taking advantage of this opportunity. For example, with the deregulation of the coal industry in 1983, a million local pits were hastily opened and non-state production came from mines which produce the more impure forms of coal.[28] Continuing shifts in the production structure will have a further detrimental effect on pollution control over the next decade. On the consumption side, rural enterprises, the small privately or cooperatively owned workshops which expanded rapidly in the second half of the 1980s, are placing great demands on rural commercial energy supplies and are too numerous and decentralized to be monitored effectively by the state.

Pricing and monitoring practices must also be adjusted along with technological improvements. China, like most command economies, historically has kept the price of industrial inputs low to effect a transfer of capital from the agricultural sector to the industrial sector. Despite chronic shortages, electricity prices have been kept low in comparison with other countries in the region—a kilowatt hour costing three cents in China, six cents in Indonesia, eight cents in South Korea, and seventeen cents in Japan. Bringing energy prices to scarcity levels would encourage conservation, but the disruptive ripple effects of such a change are feared by economic planners and cadres alike. In the summer of 1988, Zhao Ziyang did advocate the removal of all price controls within four or five years, but his views were completely rejected by the Chinese Communist Party's Thirteenth Central Committee Third Plenum that autumn, and Zhao was forced to make a self-criticism.[29]

The question of pollution monitoring or using monetary incentives is quite complicated since the Chinese economy is in a (perhaps permanent) stage of transition from plan to market. Far from an omnipotent state, China's central government is handicapped in attempts to comprehensively monitor the economy by a lack of funds, technology, and communications infrastructure. This weakness of the central organs of the state was a major reason that economic planning was never as comprehensive in China as in the former Soviet Union. China does not as yet have the personnel to carry out even basic environmental policing of energy and industrial enterprises. The heavily state owned energy sector has also been relatively impervious to fines or effluent fees. In levying a pollution fine, the government is, in effect, fining itself, and a managers have little incentive to endanger their ability to meet more important production quotas by complying with pollution regulations. Only a system where managers themselves are penalized for pollution and waste can make regulation feasible.

The possibility of extending the scope of the polluter pays principle depends on China's ability to bring to the industrial structure the type of incentive-based reforms that have been successful in the rural sector. While market mechanisms theoretically may be a more efficient means of pollution control in a nation with as weak a supervisory system as China's, it must be recognized that such a means also entails a considerable amount of often sophisticated monitoring to measure effluents and calculate treatment charges. So-called production charges could be a more simply implemented partial solution. Taxes on pollution generating commodities such as lubricants, fertilizers, and disposable containers have been used in a number of European countries, but, again, success hinges upon the willingness of planners to accept a radical restructuring of relative prices. China, like other command economies, has a shortage of numerous commodities, causing demand to be price inelastic and compromising the effectiveness of price mechanisms.[30] Market strategies cannot therefore be considered a panacea, and a system of combined regulation and economic measures must be built.

The Non-Commercial Sector: Environmental Effects

Non-commercial energy accounts for thirty percent of the Chinese total, and its use in the rural energy sector is the source of some of China's most intractable environmental problems. Since extensive provision of commercial energy from centralized sources of coal, oil, or large-scale hydroelectric power is financially and logistically impossible during the next decade, the solution to these problems can only be found in improved management of local sources. Rural areas consume just under half of China's total energy, around a quarter of which is for agricultural and enterprise use, mostly locally produced coal and small-scale hydropower for pumping irrigation water. Over eighty percent of the remaining household use is supplied by biomass, mostly firewood, sticks, grasses, and crop residues.

As of 1980, seventy million of China's 170 million rural households had severe yearly fuel shortages of over three months.[31] While legislation may be effective in regulating industrial polluters, no amount of legislation can stop tens of millions of impoverished peasants from clearing hills of trees, brush, grass, and roots for fuel. These practices, as in many other poor countries, have set in motion an ecologically disastrous chain of events.

Removal of the tree cover, and, perhaps more importantly, the soil binding grass and root systems has been the prime cause of soil erosion which has plagued China. Soil nutrient loss through erosion is variously estimated at equivalent to from fifty percent to 200 percent of annual chemical fertilizer production.[32] Silt loads in Chinese rivers have also increased appreciably, and siltation in turn has directly affected energy production. Annual loss of reservoir volume in hydroelectric generating stations owing to siltation is equivalent to one-third of new reservoir volume excavated.[33] This reduces power generation and has made costly shutdowns and repairs necessary. In the Sanmenxia reservoir, unexpectedly rapid silt build-up amounting to sixty percent of reservoir capacity between 1958 and 1973 forced a major re-engineering of the project.[34] The TGP will suffer from similar problems and will not be able to store water during the rainy season when flood control is most needed because of the danger of siltation.[35] The removal of trees and their water retaining capacity combined with overzealous reclamation of lakes has made flooding more frequent: since 1949 so-called natural catastrophes have increased in frequency.[36] It has been argued, therefore, that reforestation of the upper watershed of the Changjiang would provide a natural water storage capacity equivalent to the TGP reservoir and do a better job of flood control.[37] Because of the siltation problem, reforestation must be viewed as a precondition for large downstream projects, if not a replacement.

Statistics regarding deforestation and afforestation are extremely inaccurate, but according to the conventional view, since Liberation in 1948 the Chinese have fought "a people's war of afforestation."[38] The FAO reported in 1974 that "there now exists a degree of forest consciousness that is unequaled anywhere in the world, save possibly in parts of Canada and some parts of the Nordic countries."[39] Whether or not this was ever true, the Chinese Communist Party leadership should have realized that success in this "superstructural" realm was no guarantee of results when the imperatives of economics and the relations of production were working so decisively against the environment.

Land reform in the early 1950s caused a rash of tree cutting as new owners feared policy change and old owners took their last profits.

Collectivization in forestry preceded that in agriculture because of the perception of even greater benefit in economies of scale and long term planning. Afforestation campaigns, however, were usually conducted through a mobilization method, with peasant levies conducting most of the planting. Such practices proved wholly unsuitable to forest management which requires low-intensity but continuous care of saplings, and a ten percent survival rate is often cited as typical.[40] Maintenance of existing forests was also slighted in the emphasis on quantitative targets for new planting.[41]

Prices for timber in China have lagged behind almost all of the commodities with which it competed in hilly areas, reducing economic incentives for communities to afforest on their own. This problem has continued in the post-Mao period, as the domestic quota price of timber has remained far below the price paid for imports.[42]

The Rural Sector: Environmental Reforms

Forestry was in the forefront of rural change in the post-Mao era as private hillside silviculture plots were distributed to farming families. Just as occurred during land reform, this was accompanied by a brief wave of deforestation by farmers who feared future retraction of the new policy. The contract responsibility system in forestry was introduced in 1983, with contracts of up to fifty years, compared to fifteen years in agriculture at that time. While the market-oriented policy does hold out the promise of gains similar to those made by the agricultural sector, results will not be certain until well into the 1990s. For the foreseeable future, the prospect remains of peasants forced to spend more time and effort gathering fuel from more distant locations—likely those of greater ecological fragility. With continued rural population growth, rising consumer expectations, and the new demands of rural industry, better management of wood must be supplemented by other sources. Because provision of central sources is unlikely in the near future owing to cost and transport problems, China must attempt to rehabilitate some aspects of Maoist energy policy—namely, small-scale hydropower (SSH) production and biogas.

This may at first seem to contradict the arguments for economies of scale and advanced technology, but these techniques are a necessary stopgap measure to deal with the shortages of capital, the long lead times, and, particularly, the ecological uncertainty of large hydroelectric projects.

The golden age of SSH was the latter part of the Great Proletarian Cultural Revolution (1966-76), but during the reform period SSH's share of total hydroelectric production has declined. This decline is partially because of the decommissioning of badly designed and hastily constructed dams, partly because of siltation and changed hydrological patterns, and partly because of a preoccupation with large projects. It is estimated that only ten percent of rural SSH potential has been tapped,[43] and the theoretical capacity is distributed evenly enough that over half of China's more than 2,000 counties have the potential to install 10,000 to 30,000 kilowatts.[44] Yet, over the course of the Seventh Five Year Plan, the central authorities have allowed, but not encouraged, the development of SSH in favor of larger projects which give the center more control over energy allocations.[45] Unfortunately, these projects have a tendency to fall behind schedule and are plagued by problems of cost overruns, luring appropriations bit by bit while hiding the total final cost (referred to as *"diaoyu gongcheng,"* or "fishing"). Large dams are fraught with uncertainty, since broader or unforeseen environmental effects increase exponentially with linear growth in dam size. In the meantime, even the sixty percent of Chinese villages which have electricity suffer from uncertainty of supply. On environmental grounds as well SSH projects are devoid of significant environmental impacts. Given the rapid expansion of rural enterprises since 1985, the need for SSH is particularly great. The countryside around Suzhou, for example, one of the most commercialized and rural enterprise-intensive areas in China, experienced a fifteen day power outage in the summer of 1988.[46] Provisioning with electricity where substitution is possible may be one way of preempting increased use of extremely low quality coal.

Perhaps the greatest economic loss from the disbanding of the communes has been the elimination of a successful vehicle for large-scale investment in agricultural infrastructure. Irrigation and other investments beyond the means of the household have suffered in recent years, and SSH is similarly affected. If the commune or some substitute investment conduit is not found, the possibilities for expanding SSH are slim, and reliance on destructive solid fuels will continue.

Another casualty of post-Mao reform is the biogas program, at one time heralded as a model for other poorer nations to emulate. Essentially the controlled fermentation of plant and animal matter, such as crop residues and excrement, in airtight pits, biogas doubles the energy efficiency of organic fuels producing gas for cooking and lighting, at half the cost of SSH. While pits can be constructed from local materials, they are practical only in southern China, owing to climactic constraints on their operation. Fortunately, this is the area in China that is most deficient in energy. The biogas program also reached its peak in the early 1970s, but for reasons that are not entirely clear, the program was abandoned at the time of the post-Mao transition. It is estimated that only one-third of the pits were functioning at this time.[47] Farmers may merely have been taking advantage of construction subsidies in building their pits, but more likely, biogas failed for reasons analogous to those which crippled the afforestation program: construction was on a mass mobilization basis seeking strictly quantitative goals, with little attention paid to quality, technical improvements, or maintenance. It is also possible that biogas was scarred by its association with Maoism and was too unglamorous for a place in the Four Modernizations.

Biogas warrants a reevaluation, however, if only for its environmental value. Biogas pits do not remove significant nutrients from the ecological cycle and are a means to safely dispose of parasite bearing wastes. Biogas, thus, has the ancillary effect of checking the spread of Bilharzia and other parasite diseases which have also made an unexpected comeback in the reform period.[48]

Energy Alternatives: Tentative Conclusions

We ought not to entertain fantasies of such benign technologies filling more than niches in the overall energy picture for the foreseeable future, but the Chinese government must recognize that a wide spectrum of modernizations in the energy sector are needed to best make use of poorly situated resources and minimize environmental impacts. One of the post-Mao leadership's biggest oversights in economic planning has been its failure to take into account the variety of experiences with Maoist development. In agriculture the wholesale abandonment of the commune system has resulted in a deteriorating rural infrastructure, and a new role for the commune in energy provision and environmental supervision is necessary to reduce pressure on China's forests and prevent the spread of fossil fuel dependence in the countryside. The main tension will be how to combine such structures with market oriented reforms, such as private family woodlots and specialized forestry households.

The urban sector cannot rely on such environmentally benign energy technologies as SSH and biogas: coal is here to stay as modern China's dominant energy source. The World Bank asserts that China's unique challenge is to harness coal use as a major fuel for diverse activities. Given the environmental effects of this coal use, substituting for coal use and mitigating the effects of coal use as a major fuel is a more appropriate challenge. The commercial sector will be increasingly forced to rely on modernization of technology for both increased efficiency and control of environmental degradation: better quality coal, bigger and more efficient thermal stations and industrial processes, and early installation of emission control equipment.

Returning to the question of the TGP, has "the Chinese energy industry...entered a new stage of big units, extra-high voltage and large networks" as the Seventh Five Year Plan asserts? The theoretical energy output of the TGP, eighty-four billion kilowatt hours per year, is equal to that of thirty-six million tons of coal burning,[49] and, therefore, represents a significant slowing in the growth of sulphur dioxide and other emissions. The TGP does not deserve a knee-jerk reaction: hydropower must be emphasized. Still, it seems imprudent to accept the ecological uncertainties of the mega-project when alternative merely large projects are available at comparable cost, especially given the tendency for cost overruns to increase with increasing scale. The Ertan, Longtan, and Wuqiangxi projects represent sixty percent of the projected output of TGP at proportional cost. As proponents of the Ertan project say, "we should not delay the marriages of the second daughter and the third daughter because the eldest daughter [Sanxia] has not yet been married off."

The *Sichuan Daily* of 21 September 1988 described the phenomenon of "begging with a golden bowl," referring to the fact that Sichuan, with abundant local hydropower potential and coal reserves, suffers from recurring power shortages. Blame is placed on the Beijing bureaucrats who ignore regional development because of their obsession with megaprojects. Despite its undoubted attractions, given the menu of alternatives that could fill the void, the TGP should receive a reluctant rejection. In any discussion of Chinese development it is obligatory to use the slogan "walking on two legs"—the Maoist dictum for combining modern and traditional technologies—but in the field of energy it seems that the two sectors will continue walking their own paths for some time to come. If so, what are the implications for China of the environmental and energy constraints outlined above? Will China be able to complete the Four Modernizations in some form, and what will be the effects of the energy and environmental picture on economic, spatial, and political organization?

Implications of Energy and Environmental Issues

In 1976, the year of Mao's death, the Ten Year Development Program was announced, and plans were made to construct a large number of heavy industrial megaprojects, including numerous dams and other large-scale energy installations. These plans were abandoned within two years, and came to be characterized as the last gasp of the Stalinist development mentality.[50] Still, in 1980, projections of energy needs to accompany the quadrupling of China's GNP by the year 2000, called for a commensurate quadrupling of energy supplies to 2,000 million tce, making China the world's largest consumer of energy. Assuming a relatively stable energy mix, such energy use would have disastrous consequences for the natural environment and human health. Fortunately, the assumed income elasticity of demand for energy of 1.0 was inconsistent with the experience of the OECD nations in the oil shock era from which China had been insulated by its economic autarky. From 1971 to 1976, at the end of the Maoist period, China's elasticity swelled to 2.55 as increasing inputs of energy were needed to wring output increases from a stagnating economy. From 1978 to 1981, a major conservation campaign was undertaken and elasticity fell to 0.2, and it remained at 0.6 from 1982 to 1989.[51] The majority of this saving is actually attributable to sectoral shifts in the economy from heavy industry to less energy intensive light industry and agriculture,[52] and to the results of broad directives to save energy, rather than to technological change. While easy savings have been exhausted and such savings will be harder to sustain, most planners now set a more achievable goal of 1,400 to 1,500 tce for the year 2000.[53]

Projections differ by source of energy, with Chinese planners seeing an increased share for oil and gas, while the World Bank predicts coal will become even more dominant. All agree, however, on the need for hydroelectricity to expand its share from 4.5 percent for reasons both of living standards and environmental protection.[54] Correcting the historical dominance of thermal over hydropower would have a beneficial effect, but thermal power historically has been relied on because energy planning has tended to be for catching up with demand, rather than anticipating it. With China's poor history of industrial safety, promoting a nuclear industry which has unacceptably high capital costs and environmental risks even in the West cannot be recommended.[55]

Environmental considerations also demand a continuation of the current policy of emphasizing light manufacturing and agricultural development, and recommends more emphasis on the small service sector. The necessity of price restructuring and raw material conservation, and for effluent fees to remove the economic rationale for polluting, also point the economy in the direction of greater reform.[56] Environmental considerations also rule out a return to a more closed economy since foreign technical assistance is vital to the improvement of environmental protection, and since China is relatively weak in environmental science.[57] In the long run, expansion of power grids and inter-regional energy flows to solve energy problems, and the coordination of pollution management to prevent excesses of pollution transfer and to promote rational distribution of environmental stresses will also work against the former policy of autarky.

Energy and environmental constraints will have a potentially great effect on China's spatial organization. With one-third of national industrial output, the arc of coastal provinces from Jiangsu to Hainan produces less than ten percent of China's energy and consumes seventeen percent.[58] The economic growth of these leading regions cannot be maintained under these circumstances, and will have to be either a transfer of energy intensive industries to better endowed regions in the interior or concentration on industries that require less energy input. Such a movement of industries goes against the trend in the reform era which reversed Mao's policy of building industrial centers in the interior, but this devolution of production would mirror the trend towards an increasing division of labor within East Asia. Just as Japan has exported its more polluting industrial processes to the NICs and the ASEAN nations, and as the NICs are attempting to export them to the ASEAN nations and China, China may seek to shift more of its pollution-prone industries from the coast to adjacent interior areas.[59] Shanghai already treats its hinterland "like a Third World country," internally exporting as much of its pollution problem as possible.[60] It appears that China's division between the coast and the interior—already differentiated in terms of technology and wealth—will become marked by levels of environmental degradation as well. The interior, already beset by substantial soil and water problems, will become increasingly polluted as well, and, unlike the more prosperous coast, it will be unable to pay for mitigating measures.

The Chinese state has, for at least 2,000 years, found it difficult to walk the line between guidance and strangulation of the economy, and the balancing of pollution regulation and market measures, or of commune planning and peasant initiative, will demand sophisticated judgement in this economically and politically fluid period of development. While the state must release its grasp on the economy by allowing prices to move to levels which encourage conservation, it must still play a supervisory role in the introduction and diffusion of technology, especially in the countryside, to promote optimal choices of scale and technological level, and for setting effluent fees and taxation measures to remove the economic rationale for polluting. Since much of the raison d'etre of the bureaucracy has been removed by the reformers, one could expect that these new duties could be pursued with some relish.

Politicization of the planning process is not unique to China, and megaprojects often have symbolic values which transcend rational calculation,[61] or may simply serve as battlefields for elite power struggles.[62] Environmental crises can also be used to further political ends, as when Zhao and Hu Yaobang capitalized on the 1981 floods to promote contract responsibility woodlots.[63] Such a policy-making atmosphere is not, however, conducive to rationality. Specifically important for environmental considerations is a lack of confidence in the permanence of policy, which causes short-term profit-taking in spheres that demand long-term management. Environmental degradation accelerated during the more disorderly periods of land reform, collectivization, and post-Mao reform. In a way, any particular policy is preferable to vacillation.

Looking at things dialectically, energy and environment in turn have a strong potential effect on political stability. In 1978, a Chinese government document spoke of "numerous cases of farmers occupying or even sabotaging factories that threatened their water supply or fields with poisonous emissions. In Qinhuangdao...farmers dammed the outflow of a chemical factory such that effluent flooded the factory grounds."[64] In 1979, Shanghai reported 339 "confrontational" incidents arising from environmental grievances.[65] Even in authoritarian, one-party states such as Taiwan, environmental issues have served to mobilize or focus dissent,[66] and it was the environmental issue which cracked the apparently monolithic authority of the Japanese Liberal Democratic Party at the end of the 1960s and gave rise to large, popular opposition movements.[67] Political upheaval such as Japan experienced "could shatter a less stable developing country."[68] Environmental conflict has an added dimension in China, as a number of environmental grievances originate with already restless minority groups. The Chinese leadership must certainly hope to avoid the emergence of an aggressive independent environmental movement, and for the sake of its own preservation must attempt to preempt or co-opt any such impulse.

It is evident from the TGP controversy that the environment is already shaping discussion about economic development in general. The evaluation of energy policy, in particular, needs to be conducted within a context that is more conscious of environmental issues. The Chinese government must realize that environmental considerations, and their effect of energy, are just as much a limit to growth as transportation infrastructure or energy shortage itself. Portions of academia and the political elite in China recognize that economic development is not threatened by environmentalism; rather, long-term economic and political stability are threatened by a lack of attention to the environment. Only by grasping this dialectical relationship and carefully managing the environment can the Four Modernizations succeed. Failure to do so will result in flood, drought, and blight: the familiar signs that the rulers have lost the Mandate of Heaven.

Notes

1.   An earlier version of this paper was published in *Chinese Geography and Environment*, Vol. 3, No. 3, 1990, pp. 3-23. I would like to thank Joseph Whitney and Linda Hershkovitz for their kind assistance.

2.   Bernhard Glaeser, ed., *Umweltpolitik in China* (Bochum: Brockmeyer, 1983), and Vaclav Smil, *The Bad Earth* (Armonk, NY: M.E. Sharp, 1984) are two of the earlier revisionist views.

3.   Various estimates of annual damage from environmental degradation are in the range of ten percent of GNP.

4. Thane Gustaffson, *Reform in Soviet Politics* (New York: Cambridge University Press, 1981).

5. World Resources Institute, *World Resources 1990-91* (New York: Oxford University Press, 1990).

6. Sun Jingzhi, *Economic Geography of China* (Hong Kong: Oxford University Press, 1981).

7. Xin Dingguo, "The present and long-term energy strategy of China," in James P. Dorian and David G. Frindley, eds., *China's Energy and Mineral Industries: Current Perspectives*, pp. 43-54 (Boulder: Westview, 1988).

8. If we assign a value of 100 to China's average per capita energy reserves, regional values would be as follows: north 416, northwest 146, southwest 167, northeast 40, east 22, south 19; Sun, *Economic Geography of China*.

9. Vaclav Smil and William Knowland, *Energy in the Developing World* (Oxford: Oxford University Press, 1980); Tatsu Kambara, "China's energy development during the readjustment and prospects for the future," in *China Quarterly*, No. 100, 1984, pp. 762-782.

10. State Statistical Bureau, *1990 China Statistical Yearbook*. Coal as a proportion of total energy production declined steadily in the decades following the Second World War, from over ninety percent in the 1950s, to eighty percent in the 1960s, and to a low of 69.5 percent in 1976. In 1981, the share of coal began to rise slowly again, in keeping with world trends, settling in the mid-1980s in the seventy-two percent to seventy-three percent range. Coal consumption has shown a similar trajectory as the overall patterns of energy production and consumption diverge only where crude oil export becomes significant in the late 1970s. Coal's share of consumption remains slightly above its share of production.

11. For a history of some of these problems, such as deforestation and hasty reclamation of lakes for farmland, see Yi-fu Tuan, *China* (New York: Aldine, 1969).

12. World Resources Institute, *World Resources 1990-91*.

13. Toufiq A. Siddiq and Zheng Ching Xian, "Ambient air quality standards in China," in *Environmental Management*, Vol. 8, No. 6, 1984, pp. 473-479.

14. Wolfgang Kinzelbach, "Energie und umwelt in China," in Glaeser, *Umweltpolitik in China*, pp. 303-324.

15. Kinzelbach, "Energie und umwelt in China"; Wolfgang Kinzelbach, "China's energy and the environment," in *Environmental Management*, Vol. 7, 1983, pp. 310-319; Lester Ross, *Environmental Policy in China* (Bloomington, IN: Indiana University Press, 1988); and Smil, *The Bad Earth*.

16. Xin, "The present and long-term energy strategy of China."

17. World Resources Institute, *World Resources 1990-91*.

18. Vaclav Smil, "China and Japan in the new energy era," in Peter Nemetz, ed., *The Pacific Rim: Investment, Development and Trade* (Vancouver: University of British Columbia Press, 1987).

19. Huang Yuanjun and Zhao Zhongxing, "Environmental pollution and control measures in China," in *Ambio*, Vol. 16, 1987, pp. 473-479.

20. Vaclav Smil, "A technological future for China," in *Ambio*, Vol. 8, 1979, pp. 94-101.

21. Mahesh Chaturvari, "'Second best' technology as first choice," *Ambio*, Vol. 8, 1979, pp. 71-81; and Smil, "A technological future for China."

22. Peter J. Poole, "China threatened by Japan's old pollution strategies," in *Far Eastern Economic Review*, 23 June 1988, pp. 78-79.

23. Thomas Rawski, *China's Transition to Industrialism* (Ann Arbor, MI: University of Michigan Press, 1980).

24. Kinzelbach, "Energie und umwelt in China."

25. Kinzelbach, "Energie und umwelt in China"; and Smil, "A technological future for China."

26. One hundred billion kilowatt hours is about a quarter of late 1980s production. See Smil, "A technological future for China," for an analogous calculation.

27. See Robert M. Wirtschafter and Ed Shih, "Decentralization of China's electricity sector: Is small beautiful?" in *World Development*, Vol. 18, No. 4, 1990, pp. 505-512.

28. Xin, "The present and long-term energy strategy of China," p. 50.

29. Lowell Dittmer, "China in 1989: The Crisis of incomplete reform," in *Asian Survey*, Vol. 10, No. 1, 1990, pp. 1-25.

30. Ueta Kazuhiro, "Dilemmas in pollution control policy in contemporary China," in *Kyoto University Economic Review*, Vol. 58, No. 2, pp. 51-69.

31. Sun, *Economic Geography of China*.

32. China New Analysis, 15 February 1989; Ma Hong, "Strive to improve our country's environmental protection work," in *Chinese Law and Government*, Vol. 19, No. 1, 1986, pp. 12-29; and Li Jinchang, Kong Fangwen, He Naihui, and Lester Ross, "Price and policy: The keys to revamping China's forestry industry," in Robert Repetto and Malcolm Gillis, eds., *Public Policies and the Misuse of Forest Resources* (New York: Cambridge University Press, 1988).

33. Kinzelbach, "Energie und umwelt in China."

34. Smil, *The Bad Earth*.

35. Luk Shiu Hung and Joseph Whitney, "Editor's introduction," in *Chinese Geography and Environment*, Vol. 1, No. 4, 1988, pp. 3-25.

36. China News Analysis, 15 February 1989. One of China's most serious floods in this century took place in 1991.

37. Phillip M. Fearnside, "China's Three Gorges Dam: Fatal project or step towards modernization," in *World Development*, Vol. 16, No. 5, 1988, pp. 615-630.

38. Robert Taylor, *Rural Energy Development in China* (Baltimore: Johns Hopkins University Press, 1981).

39. Taylor, *Rural Energy Development in China*.

40. Lester Ross, "The implementation of environmental policy in China," in *Administration and Society*, Vol. 15, No. 4, 1984, pp. 489-516.

41. Ross, *Environmental Policy in China*.

42. Sizeable increases in the above plan price of timber have in some regions brought the spot price in line with scarcity value; Li, *et al*, "Price and policy."

43. Clifton W. Pannel and Joseph Ma, *China: The Geography of Development and Modernization* (London: Edward Arnold, 1983); and Xin, "The present and long-term energy strategy of China," p. 52.

44. Sun, *Economic Geography of China.*

45. Stephen R. Cain and W.A. Kerr, "China's changing development strategy: The case of rural electrification," in *Canadian Journal of Development Studies*, Vol. 8, No. 1, 1987, pp. 81-96.

46. China News Analysis, 1 December 1988.

47. Rudolph Wagner, "Biogasnutzung in ländlichen und städischen regionen," in Glaeser, *Umweltpolitik in China*, pp. 365-394.

48. Wagner, "Biogasnutzung in Landlichen und Stadischen Regionen."

49. At forty percent efficiency. Compare also with the 1985 electricity shortfall of forty to fifty million kilowatt hours cited in Luk and Whitney, "Editor's introduction."

50. Cain and Kerr, "China's changing development strategy."

51. State Statistical Bureau, *1990 China Statistical Yearbook.*

52. Smil, *The Bad Earth*; Kambara, "China's energy development during the readjustment."

53. Xin, "The present and long-term energy strategy of China"; Vaclav Smil, *Energy in China's Modernization: Advances and Limitations* (Armonk, NY: M.E. Sharp, 1988).

54. Current per capita household consumption of electricity is 31.5 kilowatt hours a year, the equivalent of less than an hour per day of one 100 watt bulb.

55. Xin, "The present and long-term energy strategy of China"; China News Analysis, 15 March 1989.

56. Ross, *Environmental Policy in China.*

57. Ma, "Strive to improve our country's environmental protection work."

58. China New Analysis, 1 February 1989.

59. Mark A. McDowell, "Development and the environment in ASEAN," in *Pacific Affairs*, Vol. 62, No. 3, 1989, pp. 307-329.

60. Kinzelbach, "Probleme und terderzen in der praxis des umweltschutzes in China," in Glaeser, *Umweltpolitik in China*, pp. 397-415.

61. Luk and Whitney, "Editor's introduction." See John Waterbury, *Hydropolitics of the Nile Valley* (Syracuse, NY: Syracuse University Press, 1979), for a detailed discussion of the Aswan Dam case.

62. There has been speculation that economic rationalist Zhao Ziyang was opposed to the TGP, while the economic throwback and former minister of electric power Li Peng was in favor, but this view is criticized in some detail in Kenneth Lieberthan and Michel Oksenberg, *Policy Making in China: Leaders, Structures and Processes*, pp. 269-338 (Princeton, NJ: Princeton University Press, 1988).

63. Ross, *Environmental Policy in China.*

64. Kinzelbach, "Probleme und terderzen in der praxis."

65. Ross, *Environmental Policy in China.*

66. Margaret Scott, "Activists who fight clean," in *Far Eastern Economic Review*, 25 February 1988, pp. 44-45.

67. Gesine Foljanty-Jost, *Kommunale Umweltpolitik in Japan: Alternativen zur Rechtformlichen Steuerung* (Hamburg: Institut für Asienkunde, 1988).

68. Poole, "China threatened by Japan's old pollution strategies."

10

Japan and the Environmental Degradation of the Philippines

Rene E. Ofreneo

The economic expansion of Japan at the expense of resource-rich developing countries is well-known. The Philippines is a good example of this one-sided arrangement.

The purpose of this chapter is three-fold: first, to outline the politico-economic factors which helped shape the kind of resource relationship that developed between the Philippines and Japan; second, to show that this relationship has been changing based on the structural adjustments that both countries have been adopting; and third, to document the harm done to the Philippine environment as a result of this kind of resource relationship between the two countries.

This chapter also approaches the above tasks in a historical manner, that is, it looks into the evolving patterns of resource relationship as products of the politico-economic processes dominant at each historical period. In this regard, this chapter sees two distinct historical stages in the resource relationship between the two countries. The first was in the 1950s and 1960s, when Japan was rapidly re-establishing and expanding its industries and emerging as a dominant force in the global economy. The second stage was in the 1970s and 1980s, when Japan started restructuring its economy in favor of more sophisticated production processes at home simultaneous with greater overseas investments and relocation in less developed countries of its labor-intensive, energy-intensive, and pollution-intensive industries.

First Stage: War Reparations and Japan's Economic Expansion

Like in the other East Asian countries, Japan had difficulties normalizing trade relations with the Philippines, which suffered heavily in the hands of the Japanese during the war, after World War II. In fact, up to the early 1970s, the Philippine Congress was debating on how to conduct foreign relations with Japan and how to treat the issue of war reparations from that country. Ironically, the war reparations themselves became the very instruments by which Japan was able to re-establish its economic foothold in the Philippines.[1] The war reparations were also used by Japan to secure the scarce natural resources its growing industries needed.

Politics and Economics of Reparations

Japan started paying war damages to the Philippines only in 1956. These reparations, which lasted till 1976, totalled US$550 million and fell far short of the original Philippine government demand of US$8 billion.[2]

Why did the reparations start only a decade after the war? And why was the total amount much less than what the Philippines was demanding? The answer lies in the dramatic transformation of the American perception of what should be the role of Japan in the post-World War II era.

As is well-known, the United States, the lead caretaker of defeated Japan, decided to rehabilitate and even strengthen Japan based on American Cold War calculations that a modern but pro-American Japan could stem the communist tide in Asia and serve as America's "junior partner" in the region. Thus, to the bitter amazement of the Philippines and other countries victimized by Japan's war aggression, the United States, instead of punishing Japan and forcing the latter to comply with their collective demand for the relocation of some of Japan's industries to the claimant countries as indemnities and as insurance against any future Japanese re-armament, speeded up the rehabilitation of Japan. While the United States initially showed some sympathy toward the reparations demands of the Philippines and other Asian countries, it eventually took a firmer stand against reparations so as not to unduly weaken Japan.

In 1950, John P. Dulles, who was greatly agitated by the developments in China and the Korean peninsula, drafted a "peace treaty" between Japan and America's Asian allies which omitted any demands for reparations. Dulles only acquiesced to some form of reparations when she realized that:

> Japan has a population not now fully employed and it has industrial capacity not now fully employed and both of these aspects of unemployment are caused by lack of raw materials. These however are possessed in goodly

measure by the countries which were overrun by Japan's armed aggression. If these war-devastated countries send to Japan the raw materials which many of them have in abundance, the Japanese could process them for the creditor countries and by these services, freely given, provide appreciable reparations.[3]

The reparations that Dulles had in mind were "service reparations" or labor and technical services rendered by Japanese experts and skilled workers in the claimant countries. Eventually, through arm-twisting and the anti-communist hysteria, the United States succeeded in forcing the Philippines and other aggrieved countries to sign a "peace treaty" with Japan in 1951, with the undefined "service reparations" serving as some kind of a sweetener.[4]

This treaty paved the way for bilateral negotiations between the Philippines and Japan, which was then granted its political sovereignty. Trade between the two countries was also formally resumed. However, fuller commercial ties would not be established until 1956, when a new reparations agreement was concluded, which the Philippine Senate required as the price for its ratification of the 1951 peace treaty with Japan.[5]

It took the two countries five years to hammer out a new reparations agreement, despite American pressures, because of the strong anti-Japanese sentiments in the Philippines and the "nationalistic" posturing of a number of Filipino legislators, who were demanding better reparations terms. The new agreement provided for war payments worth US$550 million in the form of capital goods and services, payable within a period of twenty years, and economic development loans worth US$250 million. A Filipino senator, who was critical of some parts of the agreement, prophetically warned that: "The resultant effect of the set-up will be to peg Philippine industries to the raw-material needs of Japan, a situation which will largely contribute to the highly probable reduction of the Philippines into an economic vassal of Japan."[6]

With the agreement, trade between the two countries immediately shot up. The "capital goods reparations"—mainly in the form of old machinery, equipment, and chemical products—helped support the import substitution program of the Philippine government, especially in the textile industry. However, it also helped carve out a special market for Japanese products, largely industrial raw materials, in the Philippines. As to Japanese imports, the bulk were raw materials as predicted by Dulles and the critics of the reparations agreement. From the mid-1950s (when the reparations agreement was signed) up to the 1970s, two major raw material-based industries grew literally around the Japanese market. These are the logging and mining industries, whose unrestricted operations have led to so much degradation of the Philippine environment.

Rape of the Philippine Forests

Philippine forest lands are badly balding. A 1989 report of the Department of Environment and Natural Resources graphically captures the extent of deforestation in the following words:

> The total land area of the Philippines is 30 million hectares. About 50 percent or 15 million hectares are classified as forest land while 47.10 percent or 14.12 million hectares are classified as alienable...However, only about 6.5 million hectares of total forest lands are still forested, with an estimated less than a million hectares classified as primary or virgin forests. According to consolidated National Forest Resources Inventory—Systems Pour Space L'observacio de la Terre (NFRI-SPOT) data, our forests continue to be denuded at the rate of about 119,000 hectares per year...[7]

Who are the culprits in the massive denudation of the Philippine forests? There are many but the leading candidates are the past and present participants in the Japan-oriented logging industry. In the early 1950s, forest products accounted for five percent of Philippine exports; in 1966-68, their share increased to twenty-seven percent.[8] It was only in the last decade that Japanese importations of Philippine timber declined, mainly because the "harvestable" forests had literally disappeared. In 1966, the Philippine share of the Japanese log market was as much as fifty-three percent; in 1976, it was down to nineteen percent.[9]

The participants in the Japan-oriented logging industry include: the Japanese government, which encouraged the flow of imported timber in support of Japan's domestic industries; the Japanese importers who helped finance the operations of a number of Filipino logging companies; Filipino politicians and bureaucrats who became instant millionaires through their powers to allocate timber licenses and facilitate exportations; and the timber concessionaires themselves. A recent study shows that the Japanese trading houses helped organize the operations of some of the Filipino logging companies by advancing huge loans for the purchase of logging equipment.[10] There are also indications that Philippine exporters, with the connivance of the Japanese government, have been under-reporting their log exports to Japan as reflected in the discrepancies in the official Japanese and Philippine statistics on the Japanese log imports from the Philippines (see Table 10.1).

For the Filipino participants, consisting mostly of local politicians and bureaucrats, participation in the industry was a surefire formula for economic success. Forest authorities were charging a nominal fee of P30 per cubic meter of harvest. Very little forest management and reforestation was ever conducted. Concessionaires usually "cut and run," which explains

TABLE 10.1 Smuggling of Philippine Logs to Japan (cu. m.)

Year	Phil. figures	Japan's figures	Difference
1976	671,605	1,902,118	1,230,513
1977	1,519,315	1,737,917	218,602
1978	1,629,348	1,805,292	175,944
1979	892,576	1,400,328	597,752
1980	714,541	1,116,256	451,715
1981	705,437	1,467,033	761,596
1982	752,408	1,445,785	693,377
1983	786,036	705,614	(80,422)
1984	845,969	1,011,634	165,665
1985	454,336	558,524	104,188

Source: Haribon Foundation, 1989.

why many concessionaires abandoned their concessions in less than ten years even when they are allowed to hold on to their concession up to twenty-five years, renewable for another twenty-five years. Thus, while there were 470 concessionaires, each of which had the right to harvest 25,000 hectares of forests, in the 1960s and 1970s, only 147 have remained today. The most common and most serious breaches of the loggers were non-compliance with the norms of selective logging and the obligations to reforest logged-over areas. Once cleared, the open forest lands become attractive to the slash-and-burn farmers, who complete the process of forest destruction.[11]

What is the environmental impact of the widespread denudation of Philippine forests? An Asian Development Bank report sums it up:

Soil erosion has become widespread in the uplands and lowlands alike, and is estimated at an equivalent of one meter deep material over 100,000 ha. of land per year, or about one billion cubic meter of material every year. About 20 percent of this material is deposited in water channels, rivers and reservoirs. As a result, vast areas of forest and agricultural lands have been rendered unsuitable for growing trees or for crop farming, or the productivity of such lands has been considerably reduced. Sedimentation in rivers, reservoirs and irrigation canals has increased the threat to supply of water for agriculture and industry and even for domestic consumption. Nineteen out of the country's 57 watershed areas have already been declared to be in a critical state of degradation. Disappearance of forest and other vegetative cover has also affected watershed area of a small number of small rivers and streams on which municipal water supply depends. Flash floods have become frequent, adding to the risk of annual crop agriculture. Several species of flora and fauna have become endangered or extinct, and the actual loss of genetic material destroyed along with the natural habitats will remain unknown forever. It is feared that the environmental crisis would continue to worsen as an increasing population becomes compelled to eke out a living at the expense of nature and the generations to follow, causing, paradoxically though, a perpetuation of the misery of the rural population.[12]

Unfortunately, the Asian Development Bank was not as forthcoming in identifying who are the major culprits responsible for this grim situation.

Pollution and the Mining Industry

Like the logging industry, the mining industry in the post-war period developed around the needs of Japan. Thus, the focus of mineral exploration and development in the 1950s and 1960s was on copper and iron, two metals highly valued by Japanese industry. To promote copper exploration and development in the Philippines, Japanese smelters and traders concluded long-term contracts to assure stable prices for local producers and advanced loans in exchange for copper concentrate exports. For instance, Atlas Consolidated, the largest copper mining company in the Philippines and the biggest copper producer in East Asia, got started through a series of loans advanced by Mitsubishi Metal Mining in exchange for future shipments of copper concentrates.[13] Because of the rapid development of the copper ore mining industry, the Philippines was able to corner half of the Japanese imports of copper concentrates in the second half of the 1950s. From then on, however, the Philippine share shrank to one-third of the total as the Japanese smelters and refiners kept on expanding their capacity and as they tried to get their ores from other sources,[14] obviously to maintain greater bargaining leverage vis-à-vis the copper producers and achieve a higher degree of raw material security.

In the 1960s and 1970s, Japan emerged as the world's largest exporter of copper although its own copper mining output is minuscule.[15] This is due, of course, to its success in securing raw copper from countries like the Philippines, which had failed to develop its own smelting, refining and copper manufacturing plants.

The industrial success of Japan in copper production has a huge environmental cost, a great part of which is borne by the copper-ore producing nations like the Philippines. First, these countries lose resources which are non-renewable or which take millions of years to form, although continuing advances in technology (prospecting and processing) tend to expand the estimates of the remaining mines. Second, copper ore mining is destructive of the environment, whether done underground or open pit. Copper mining requires large land areas in both the exploratory and development phases. Forests, sometimes located in critical watershed areas, have to be cleared. The explosions and the disemboweling of mountains affect the natural ecosystems. The open-pit operations send up clouds of dust which destroy plants and lungs for kilometers around. The milling and concentrating activities emit pollutants in the air. The mine wastes or tailings made up of crushed rocks, chemicals, and traces of metals, when recklessly dumped, lead to siltation as well as poisoning of rivers, rice fields, and coastal areas, killing in the process plants, fish, and other organic materials.

All these destructive effects on the environment are observable in the Philippine mining industry. Also, Philippine mining companies double as logging companies. They apply for concessions to cut tress which they can use in mine workings and other related activities. Also, naturally, they clear the trees in their mine sites. Thus, mining companies contribute to forest denudation. In some cases, the establishment of mines becomes an occasion for the uprooting of tribal or cultural communities from their ancestral lands.

In 1977, a Filipino geographer wrote:

Twenty four active mining firms are presently discharging approximately 140,000 tons of cyanide, acid, alkaline, salt, heavy metal-containing mine tailings daily into 8 major drainage systems of the country and affecting approximately 190,000 hectares of cultivated lands. The Baguio mining district alone is discharging about 27,000 tons of waste materials daily into the lowlands and these are affecting agricultural lands as well as the fisheries industry as far away as Pangasinan and La Union. Unfortunately, the cost of these damages is difficult to estimate but the monetary value doubtless can easily run into millions of pesos.[16]

The geographer failed to mention that the biggest polluters among the mining companies are the giant copper firms. The operations of Marcopper in Marinduque island has been the object of complaints of those in the fishing and other sectors because the tailings it disgorges at Calancan Bay are killing the bay. In 1989, Marcopper had to temporarily suspend its operations because of the mounting protests of the people of Marinduque. There are also continuing complaints about the tailings dumped by Philex, Benguet Exploration, and Benguet Consolidated which find their way in the river systems of northern Luzon; the tailings of Marinduque Mines (which suspended operations in the 1980s) polluted the Sipalay River in Negros; and the tailings of the Apex, North Davao, and Sabena mines have turned the Masara River and Amacan Creek of Davao Oriental clayish white.

However, the most massive case of pollution is associated with the operations of Atlas, the copper giant. A 1981 mining study reported that

Atlas dumped some seventy-five million tons of tailings between 1955 and 1971 into the Sapangdaku River, resulting in the massive siltation and poisoning of the river, which, in turn, led to frequent flooding of the nearby villages and decline in the rice and corn harvests of farmers. Because of the numerous complaints, Atlas was forced to build a forty-six kilometer tailings pipe, which conveys its wastes off the coast of Tanon Strait in Cebu. In 1980, or barely ten years after the pipe's construction, a team of researchers from Silliman University surveyed the water area near the pipeline and found fourteen gastropod species and sixteen pelecypod species all dead! The bottom of the sea was heavily silted over a wide area, coral reefs were dead, and visibility was less than a meter.[17]

As to iron ore mining, the lifetime of the Philippine iron mines is relatively short simply because they are easily mined out. In the 1950s, the pre-war Philippine Iron Mines, which catered to Japan in the 1930s, was revived only to be closed down in the 1960s when the mineable ores ran out, creating economic dislocations in the community that developed around the company in Paracale, Camarines Norte.

The most destructive iron ore mining activity took place in the 1970s, when Japanese importers financed beach sand iron ore mining along the Ilocos and Leyte coastlines. From the mining point of view, this operation is very economical because it simply requires the shoveling of pulverized iron ore concentrated in certain beaches. But the operation is a frontal attack on the natural beauty of beaches and the coastal areas. In 1976, the Secretary of Natural Resources suspended beach sand iron ore mining on the following grounds: "(a) alarming extraction of magnetite, silica and concrete aggregates and decorative wash-out stones; (b) pollution of domestic water supply and farm lands caused by salt water as a result of the mining operations; (c) failure of mining operators/permittees to restore and rehabilitate completely the mined areas; and (d) disturbance of ecology as a result of the reported recession of the shoreline and erosion of the beaches covered by the said mining operations."[18] What the secretary failed to mention was that the affected areas—Ilocos and Leyte—happened to be the home regions of President Ferdinand Marcos and First Lady Imelda Marcos, respectively.

Second Stage: Economic Restructuring and Industrial Relocation

In 1970, the civil service chief of the Japanese Ministry of International Trade and Industry (MITI), in a dramatic speech before the OECD Industrial Committee, said:

Industrialization in developing countries will stimulate competitive relations in the markets of advanced nations in products with a low degree of processing. As a result, the confrontation between free trade and protectionism will grow more intense. The solution of the problem is to be found, according to Japan's economic logic, in progressively giving away industries to other countries much as a big brother gives away his outgrown clothes to his younger brother. In this way, a country's own industries become more sophisticated.

A solution to the North-South problem depends not only on internal development for developing nations, but also on giving them fair opportunities in the area of trade. To do this, the advanced nations must plan for sophistication of their industrial structures and open their market for unsophisticated merchandise as well as offer aid in the form of funds and technology.[19]

This big-brother-little-brother thesis is nothing but an advanced appreciation by the Japanese industrial planners of the importance of restructuring Japan's economy based on the emerging and evolving international division of labor, which sees Japan specializing in high-technology, research-intensive, and knowledge-intensive industries, on one hand, and moving out of the low-technology, labor-intensive, energy-intensive, and pollution-intensive industries, on the other hand. The latter industries are what the "big brother" Japan is now relinquishing to "little brothers" in the developing world. Such brotherly posturing, however, was grounded on less-than-brotherly concerns: the rising cost of Japanese labor which was pricing Japan out of competition in labor-intensive and low-technology manufacturing, the continuing anxieties over Japan's security in energy and raw materials, and the rising cost of maintaining energy-intensive and heavy-polluting plants in a more environmentally-conscious Japan.[20]

ADB and the Twin Programs of Agricultural Modernization and Outward-looking Industrial Development

For the big-brother-little-brother Japanese vision of a division of labor to materialize, Japan would, of course, need the cooperation of the other actors, especially the "little brothers" themselves, who would allow the transfer/relocation of the low-technology, labor-intensive, and energy-intensive industries through the medium of foreign investments. In short, this would require the economy of the little brothers to be not only hospitable to foreign capital but also to be organized for receiving the "outgrown clothes" or dated industries of big brother.

In this regard, the Asian Development Bank (ADB), the regional clone of the World Bank, has played a major supporting role in the propagation of the Japanese industrial vision in Asia, intellectually and organizationally. On the intellectual plane, the ADB tried to influence the development thrusts of developing Asian countries in the decade of the 1970s by commissioning high-profile studies of developmental economists like Hla Myint, Theodore W. Shultz, and Kazushi Ohkawa.[21] In their works, Myint and company simply tried to reinforce the sequential or gradualist approach to growth and development favored by the Japanese, arguing on the need *on the industrial front* to reduce mass unemployment through the promotion of a labor-intensive, outward-looking industrial strategy and *on the agricultural front* to reduce mass hunger through the mobilization of the countryside or the modernization of the technical base of traditional agriculture.

In a way, the development philosophy being espoused by the ADB echoes that of the World Bank and other development economists supported by the aid givers such as Gustav Ranis of Yale University's Economic Growth Center. In agriculture, the ADB approach dovetails neatly with the World Bank-favored Green Revolution based on the modern rice varieties produced by the International Rice Research Institute. On the operational level, the ADB mobilized its credit resources in the last two decades in support of the agricultural modernization and shift to the outward-looking industrial regime of member countries, especially in the building of irrigation projects required by the Green Revolution and the modernization of ports in support of a more vigorous export program.

Marcos and the Japanese

The Japanese never denied that they had very close relations with the administration of President Ferdinand E. Marcos. Records show that it was under the administration of Mr. Marcos that Japanese economic ties were fully normalized, resulting in a flood of Japanese investments and loans that nearly overshadowed those of the United States.

A facilitating factor in this full normalization of economic ties was the declaration by Marcos of martial law in 1972. According to Japanese Ambassador to the Philippines Toshio Urabe, martial law helped Japanese investment in the Philippines for three reasons: first, it paved the way for the ratification of the Treaty of Amity, Commerce, and Navigation between the two countries, which had been languishing for nearly a decade in the Philippine Senate; second, martial law helped improve the "peace and order situation"; and third, the martial law government created a new climate favorable to foreign investors.[22]

With regards to the investment climate, martial law not only reduced the restrictions on foreign investments with the fuller implementation of the Foreign Business Regulations Act (RA 5455) and the Export Incentives Act (RA 6135), but also, and more importantly, it put an end to the debates on the economic directions the government should pursue. In the pre-martial law period, an increasingly nationalistic Congress, through the Congressional Economic Planning Office, was arguing for greater economic nationalism through the Filipinization of certain industries and stronger support measures for Filipino capital; on the other hand, the increasingly powerful technocrats advising the executive branch were ardently committed to the dismantling of import substitution, greater participation in the global market through labor-intensive manufacturing, and faster propagation of the Green Revolution, objectives that both the ADB and the World Bank were strongly advocating as well. With martial law, the debate ended in favor of the technocrats.[23]

In 1972-76, most of the Japanese investments went into agribusiness firms, fishery ventures, food manufacturing, mining, wood processing, textile manufacture, and chemicals. The single biggest investment during this period, accounting for nearly forty-five percent of total Japanese investment, went into the establishment of the Kawasaki Sintering Plant.[24] Clearly, the pattern of Japanese investments reflects their priority concerns: securing food, raw materials, and energy from abroad and reducing pollution at home. The same pattern can be seen in the credit assistance given by the ADB to the Philippines.[25]

The Ecological Dimensions of Japanese-Led/Japanese-Supported Projects

With the intensification and diversification of Japanese involvement in the Philippine economy, the Japanese contribution, direct or indirect, to the environmental degradation of the Philippines has also deepened and taken on new dimensions. The following is only a brief outline of some of the new features in the environment/resource relationship between the two countries.

(1) *Propagation of the Green Revolution.* It is now well established in the scientific community that the chemical-based Green Revolution propagated by the World Bank-ADB-IRRI combine puts stress on the ecosystem, especially through the intensive use of chemical fertilizer and pesticides which have already eliminated the edible snails and other organic matters that could be found in the pre-Green Revolution rice fields. Also, chemical-based rice farming erodes the soil nutrients faster, resulting in the long run in the increasing use by farmers of greater amounts of fertilizer and other

inputs, whose prices continue to rise. Thus, despite yield increases, many farmers remain as poor as before, with some even reduced to penury.

Another ecological aspect of the Green Revolution is the establishment of a network of huge and expensive irrigation dams to insure the success of the Green Revolution. Supported by the World Bank and the ADB, many of the irrigation projects in the Philippines became controversial because of lack of careful planning and democratic consultation with the affected sectors. Thus, some of the dams led to the displacement of a number of tribal minorities, some of whom fought back violently in self-defense. Other dams, lacking the protective forest cover, are posing a danger to population centers.[26]

Ironically, in 1979, an ADB staff working paper, recognizing the socio-economic problems that have arisen in the implementation of the Green Revolution, recommended the following policy guidelines in ADB assistance: oriented to the basic needs of the poorest group; participative; economically integrative (commodities to be produced will help meet the consumption needs of low income households); environmentally sustainable (projects will preserve the productive capacity of agriculture and forestry); and cost effective or cost-reducing.[27] Despite the publication of the above recommendations, the ADB did not introduce any radical changes in its operational guidelines.[28]

Obviously, what is more important to the ADB management is the faster realization of the so-called "transformation of the technical base of agriculture" favored by Shultz and Ohkawa, which is seen as the fastest way of doubling and even tripling food production, and, thus, insuring the food security of Japan. The socio-economic situation of the farming community and the ecological balance in the rice fields are secondary.

(2) *Promotion of Export Agriculture.* Side by side with the propagation of the Green Revolution, Japanese corporate interests were actively promoting export agriculture, specifically the growing of bananas and exotic fruits for Japan. With Sumitomo teaming up with Dole, Del Monte, and United Fruits, the export-oriented banana industry developed rapidly in the 1970s, with some 25,000 hectares converted into banana plantations almost overnight. This land conversion reduced some of the settlers and members of indigenous ethnic communities into either plantation wage earners or landless rural workers.[29]

The export-oriented banana industry is a major user of agricultural chemicals. However, according to a leading Filipino entomologist, the industry is very secretive about its policy on pesticide use, to the detriment of the health and safety of the workers and families living in the plantations. The cost of pesticide use in the banana industry in 1976 was US$30 per hectare. From the observations made by the Filipino entomologist, the industry did not appear to be strict in the enforcement of safety guidelines in the handling of pesticides, especially during spray application.[30]

(3) *Fishery Modernization and Prawn Farming.* Another area of the economy where Japanese interests are strongly represented is the fishery industry. The ADB and Japanese aid agencies helped modernize the Navotas Fishing Port and Fish Market, which handles more than forty percent of the country's fish landings. A Japanese firm, Toyo Konsetsu Co., reclaimed the offshore land in Navotas and built the port. Another Japanese contribution towards the modernization of the industry was the establishment and maintenance of the Southeast Asia Fisheries Development Center (SEAFDEC), a fishery research and development center for the region.

These modernization efforts in the fishery infrastructures were complemented by Japanese investments in three dozen Filipino-Japanese joint ventures in deep-sea fishing, aquaculture, and tuna fishing. The entry of the Japanese investors was facilitated by the promulgation of Presidential Decree 704 of the Fisheries Decree of 1975, which relaxed earlier legislation providing for a minimum seventy-five percent Filipino ownership of fishing boats and which also allowed seventy-five percent of the crew to be made up of foreign nationals.[31]

The 1980s also saw the rapid growth of the Japan-oriented prawn farming. In Negros, thousands of sugar hectares were converted in the mid-1980s into prawn farms. It was only in 1989 that the craze to build prawn farms ended as a result of the collapse of prawn prices in Japan.

What are the implications of all these developments on ecology? First, the modernization and commercialization of the fisheries industry resulted in the marginalization of the small-scale fishery, found all over the country's coastlines. The big commercial fishing boats are able to lord over wide fishing areas, pushing the smaller operators to the coastal areas where over-fishing is already rampant. Worse, some Japanese and Filipino commercial fishing boats are engaged in the highly-destructive trawling and purse seine fishing, which destroys the coastal fishing grounds.

Second, the growing export orientation in the fisheries industry means less and less fishery products are being made available to the local populace despite the acute nutritional problems afflicting many Filipinos. While fishery exports help provide extra dollars to the country, they also aggravate the problem of Filipino fish consumers because the higher export prices tend to pull up domestic prices for the same products. In some cases, exportable fishery catches are simply not shared anymore with the domestic market.

Finally, in the case of the Japanese-oriented prawn farming, it has been observed that this kind of farming or fish culture requires tremendous pumping of underground water, destroys the natural water tables below the earth, and leads to the drying up of some natural springs. Saline water also seeps into the freshwater tables and mixes with the water meant for irrigation and even drinking. The lure of prawn farming has also enticed certain investors to convert mangrove areas, which are natural breeding grounds for fish, into prawn farms. Interestingly, most of the prawn farms are located in Region VI, where the Japanese-supported SEAFDEC and ADB fishery projects are also situated.

(4) *Kawasaki Sintering Project.* An integrated steel mill complex has been on the government drawing board since the end of the war. Up to now, the integration of the iron and steel industry is still an incomplete process, with the smelting and refining of iron ore constituting the vital missing links in the chain. These missing "stages" were clearly felt in the late 1960s and early 1970s, when iron ore mining was at its height, and when most of the "semi-finishing" and "finishing" steel plants were already well established, some of them even going into minor export activities. Thus, a government steel research group reported in 1972 the following observation:

> The country exports iron ores and imports finished and semi-finished steel products, principally from Japan, such that dollar expenditures on imports have greatly exceeded the country's export earnings leading to an unfavorable balance of trade. In 1971, importation of iron and steel products totalled $87 million while exports amounted to only about $9 million.[32]

The iron and steel imports of the Philippines consist mainly of bars and rods, sheets and plates, hoops and strips, structural shapes and sections, rails, tubes, pipes and fittings, and hollow mining drills. Many of these products are also processed locally by the "semi-finishing" and "finishing" steel plants which have been importing their raw materials, which consist mainly of ingots (steel cast after the smelting and refining of iron ores) and semi-finished steel products (blooms, slabs, and billets).

Ironically, an iron sintering plant now exists in the country. But this plant is not "integrated" with either the local iron mining (which is stagnant at the moment) or with the local semi-finishing and finishing plants. The sintered iron output of the plant, which is being operated by the Kawasaki Steel Corporation (KSC), is destined mainly for the steel makers in Japan; the iron ore input comes from Brazil and Australia.

Located in Villanueva, Misamis Oriental, the KSC's sintering plant is not only an oddity in the Philippine iron and steel industry, but also a highly controversial industrial project. The sintering of iron ore is an energy-intensive, pollution-causing process. The raw iron passes through an ore-sizing plant to the sintering furnace where it is mixed with limestone and other materials to form a molten lava. When the heat is turned off, the ore is reduced to pellets, which, when shipped to Japan, will pass again through a blast furnace for further processing to form pig iron, the basic raw material of steel products. A noted Filipino author essayed on the real reasons why the project was set up in the Philippines:

the Philippine Sinter Corporation, a 100 percent Japanese-owned subsidiary...is considered the single biggest Japanese investment in the country, capitalized at P478.6 million, which is roughly 44.7 percent or nearly one-half of the total amount of all Japanese investments in joint ventures here. At first blush, Japan or the investing firm, Kawasaki, appears to be genuinely interested in helping the country along the road of authentic industrialization...A closer reading of the facts, however, shows that the huge investment is only in one phase of steel-making—sintering, which, it should be noted, causes the most pollution. All other stages of steel production, particularly the processing and final phases, are still done in Japan. Clearly therefore, Japan is transferring its sintering plant here to export its problem of pollution, a raging political issue in Chiba where Kawasaki's operations have already claimed scores of lives by their heavy pollution of the atmosphere in the said area.[33]

From the beginning of its construction in 1974 up to the present, the sintering plant has been strongly opposed by environmentalists here and abroad. A Japanese, Noel Yamada, accused Kawasaki of "poisoning" Cagayan de Oro. He said that in 1978-79 there were already clear indications of the pollution of the water and air around the 144 hectare sintering site. He also deplored the sad fate of the 136 families who were displaced from the present Kawasaki site.[34]

(5) *PASAR's copper smelting and refining project*. As in the steel industry, the integration of the copper industry had long been on the drawing board, with the smelting and refining stages as the missing links. Those links are now a visible reality in the Leyte plant of the Philippine Associated Smelting and Refining Corporation (PASAR). Constructed in 1980-83, the PASAR plant now processes most of the copper concentrates produced in the country.

However, the Japanese themselves helped build the plant and they still play a key role in the operations of the plant as equity holders of PASAR, as technical consultants in the plant, as creditors in the construction of the plant, and as international marketing agents of PASAR. Marubeni coordinated the construction of the plant, with Mitsui Mining and Smelting doing the designing and engineering works. Initial funding for the project came mainly from the Export and Import Bank of Japan and a syndicate of banks led by Fuji Bank Ltd. of Japan. PASAR also entered into an agreement for three Japanese trading firms—Marubeni, Sumitomo, and C. Itoh—to handle the world marketing of the copper cathodes produced by PASAR.[35]

Thus, it seems ironic that the country that smelted most of the Philippine copper concentrates in the past became responsible for the construction of a "rival" smelter in the very country from where it has been partly sourcing its imported copper concentrates. A fuller understanding of this situation emerges, however, if we return to the big-brother-little-brother relationship advocated by MITI. Under this relationship, "wornout" industries—like smelters—may now be relocated in developing countries to reduce pollution and energy consumption in Japan.

As to the environmental impact of the PASAR project, a science professor from the Nagoya University did a survey of the sea and land around PASAR in 1990. Part of his report (written in broken English) reads as follows:

> The most polluted was the sea. By the waste water from PASAR, the sea water is strongly polluted with acid. Usually pH value of non polluted sea water is weakly alkaline, showing between 7.8 to 8.5. However, pH value of 2.8 was observed for the sea water around PASAR. This pH value is similar to that of Vinegar. In these acidic water, any living things will not alive, except some special bacteria...These surprizingly high concentration of toxic metals show clearly that the waste from PASAR has been accumulated during the past years of operation. The mud of 18,100 ppm of copper would be usable again as ore material. Highly polluted mud was distributed only at Mutlang Bay and the sea shore around PASAR. From these results, we conclude that metals in the sea was discharged from PASAR.
>
> Additionally to the research on the sea, we studied soil pollution on the land of ISABEL area...However, it seems metal concentration in the soil are not so severe to affect plant growth in these area. Very high concentration of manganese was observed for soil at every station...[36]

Conclusions

The above are major cases illustrating the environmental consequences of the kind of economic and resource relationship that the Philippines developed with Japan during the last two decades. The list of environmentally-degrading Japanese-Filipino economic projects is certainly much longer, given the fact that Japan today is a major actor in almost all key aspects of the Philippine economy as an aid giver, creditor, investor, importer, exporter, and technical consultant. Japan is now the largest source of official development assistance for the Philippines. It is also the leading financier of the "Multilateral Aid Initiative" or "Philippine Aid Plan," which was formally launched in Tokyo in 1989 to serve as a mini-Marshall Plan for the Philippines.

What then are the lessons that can be derived from the foregoing? There are a few:

(1) Japan clearly has less-than-brotherly concerns when it seeks to "transfer" a number of her labor-intensive, energy-intensive, and pollution-intensive industries to other countries as well as when it tries to project itself as champion of the "modernization" of Philippine agriculture. Among these concerns are: the need for Japanese industry to save on labor costs to remain competitive in the world market, a desire to reduce pollution in Japan, fears over Japan's long-term security in food and raw materials, and the need to economize in energy consumption in Japan.

(2) The "transfer" of industries from Japan is not complete because Japan still maintains some control—technology, marketing, and financing—over the "transferred" industries.

(3) As a result of number 2 above, the Philippines realizes limited gains in the division of labor with Japan, gains which are further reduced by the environmental and social costs of certain Philippines-Japan projects. Moreover, because of their segmented (non-integrated with the other requirements of the Philippines economy, but integrated with the global requirements of Japan's industry) and their environmentally-harmful character, these undertakings tend to be unsustainable in the medium and long run (iron ore mining, for example).

(4) Hence, there is an urgent need to review and recast Philippine-Japanese relations in a way that will make the relationship more equitable for the Philippines, allowing the Philippines to pursue growth and development in a sovereign manner and in the framework of the all-rounded and sustainable development of the entire country.

(5) Environmental degradation is not just the sin of humankind in general. It is first and foremost a consequence of the kind of politico-economic system that obtains in a given society and the kind of national leadership that a country has.

(6) The Philippines needs a comprehensive environmental and natural resources reform hand in hand with the implementation of pressing socio-economic reforms, such as provision of more urban housing and agrarian reform.

(7) If monetized, the cost of the environmental degradation in the Philippines brought about by the kind of resource relationship it has with Japan would run to tens of billions of dollars. The Philippines need not be grateful to Japan for being the biggest aid giver and investor. The Philippines may ask Japan to pay its gargantuan foreign debt and like Japan in the 1950s (with American help), assist the Philippines develop industrially in an ecologically-sustainable manner.

Notes

1. Despite American colonial rule, Japan succeeded in establishing a strong economic presence in the Philippines in the pre-war period, especially in the fishing, logging, plantation, and mining sectors. On the eve of the war, Japan was already the second trading partner of the Philippines, next to the Untied States. See Milagros C. Guerrero, *A Survey of Japanese Trade and Investment in the Philippines, 1900-1941* (Quezon City: University of the Philippines, 1967).

2. The politics and economics of war reparations are exhaustively discussed in Takushi Ohno, *War Reparations and Peace Settlement* (Manila: Solidaridad Books, 1986).

3. "Statement of J.F. Dulles, 5 September, 1951," in *Lawyers Journal*, Vol. 18, No. 1, 1952, p. 4, cited in Takushi Ohno, "US policy on Japanese war reparations, 1945-1951," in *Asian Studies*, Vol. 13, No. 3, 1975, p. 42.

4. Ohno, *War Reparations*, ch. 3.

5. Ohno, *War Reparations*, chs. 4 and 5.

6. The warning was made by Senator M. Cuenco who voted anyway for the agreement. See Ohno, *War Reparations*, p. 123.

7. Department of Environment and Natural Resources, *1989 Annual Report*, p. 15.

8. George L. Hicks and Geoffrey McNicoll, *Trade and Growth in the Philippines: An Open Dual Economy* (New York: Cornell University Press, 1971), p. 211.

9. Emma S. Vergara, "A Second Look at the Log Export Ban," in *Conservation Circular* (Los Baños: U.P. College of Forestry), Vol. 14, No. 7, 1978, pp. 3-4.

10. Francois Nectoux and Yoichi Kuroda, *Timber from the South Seas* (World-Wide Fund for Nature International Publication, 1989).

11. Masao Fujioka, *Report and Recommendations of the President to the Board of Directors on the Proposed Loans and Technical Assistance to the Republic of the Philippines for a Forestry Sector Program* (Manila: Asian Development Bank, 1988), pp. 17-18.

12. Fujioka, *Report and Recommendations*, pp. 10-11.

13. John P. McAndrew, *The Impact of Corporate Mining on Local Communities* (ARC Publication, 1983), pp. 20-21.

14. Seigo Yamamoto, "Copper forges fruitful link," in *Business Day*, Supplement on the Philippine Mining Industry, 29 May, 1987, pp. 14, 19.

15. Dorothea Mezger, *Copper in the World Economy* (New York Monthly Review Press, 1980), ch. 1.

16. P.M. Zamora, "An assessment of the environmental effects of mineral extraction and processing," paper presented in the first National Mines Research Congress, Baguio City, March 1977.

17. McAndrew, *The Impact of Corporate Mining*, pp. 56-61.

18. Agribusiness and Natural Resources Research Department, *Philippine Iron Ore Mining Industry* (Manila: Development Bank of the Philippines, 1983), p. 2.

19. Cited Ronald Dore, *Flexible Rigidities* (Stanford, CA: Stanford University Press, 1986), pp. 139-40.

20. Dore, *Flexible Rigidities*, chs. 1-2.

21. Hla Myint prepared for the ADB the book *Southeast Asia's Economy: Development Policies for the 1970s* (Middlesex, England: Penguin Books, 1971); Theodore W. Shultz and Kazushi Ohkawa co-chaired the Consultative Committee which coordinated the ADB's first *Asian Agricultural Survey* (Tokyo: University of Tokyo Press, 1968) conducted in 1967 and which guided the ADB's rural lending in the 1970s.

22. Mamoru Tsuda, *A Preliminary Study of Japanese-Filipino Joint Ventures* (Quezon City: Foundation for Nationalist Studies, 1978), pp. 6-7.

23. Alejandro Lichauco, "The international economic order and the Philippines," in Vivencio R. Jose, ed., *Mortgaging the Future* (Quezon City: Foundation for Nationalist Studies, 1982), pp. 38-47.

24. Tsuda, *Japanese-Filipino Joint Ventures*, p. 34.

25. Rene E. Ofreneo, "Modernizing the agrarian sector," in Jose, *Mortgaging the Future*, pp. 98-127.

26. Cheryl Payer documented some of these controversies in her book *The World Bank* (New York: Monthly Review Press, 1982).

27. Martin C. Evans *et al.*, *Sector Paper on Agriculture and Rural Development* (Manila: Asian Development Bank, 1979), ch. 5.

28. Robert Wihtol, *The Asian Development Bank and Rural Developments* (London: Macmillan Press, 1988), pp. 73-78.

29. Rene E. Ofreneo, *Capitalism in Philippine Agriculture* (Quezon City: Foundation for Nationalist Studies, 1980), pp. 117-118.

30. Edwin Magallona, "Pesticide use in the banana and sugar industries," in *Philippine Labor Review*, Vol. 8, No. 1, 1984, pp. 14-21.

31. Third World Studies, "Japanese interests in the Philippine fishing industry," in *AMPO*, Vol. 10, Nos. 1/2, 1978, pp. 52-60.

32. Metals Industry Research, *Primary Iron and Steel Industry of the Philippines* (Manila, 1972), p. 1.

33. Renato Constantino, "Foreword," in Tsuda, *Japanese-Filipino Joint Ventures*, p. ii.

34. Noel Yamada, in *AMPO*, Vol. 12, No. 2, 1980, pp. 70-73.

35. Shimpei Omoto, "Copper smelting and refining," paper read in the eighth Joint Conference of the Japan-Philippine Economic Cooperation Committee, March 1981, Fukuoka.

36. Masaharu Kawata, "Environmental pollution in Leyte," typescript, March 1990.

11

Debt and Environment: The Philippine Experience

Rosalinda Pineda-Ofreneo

The Philippines is saddled with a US$27.6 billion foreign debt. The country is classified as a Severely Indebted Middle-Income Country (SIMIC), ranking sixth (after Brazil, Mexico, Argentina, Venezuela, and Thailand) in the World Bank Debt Tables for 1989-1990.[1] Being a severely indebted country means being in perpetual financial hemorrhage. For the Philippines, the net resource outflow for the period 1988-1992 will be approximately a negative US$16 billion, as some US$21 billion will have been paid out in interest and principal payments while only around US$4 billion in "new money" is expected to have come in.[2]

Such a debilitating outflow robs the people of resources that could go into economic recovery and development, basic utilities and social services, and structural reforms to empower the poor and spur Filipino-oriented industrialization. What's worse, the people pay for the outflow in terms of new taxes that the government exacts to earn more revenue for debt servicing, in terms of working harder and longer while receiving less real income as a result of devaluation and inflation stemming from creditor-imposed policies, and in terms of dollars they remit as they are forced to find employment overseas as a result of debt-connected structural unemployment and underemployment. The continuing export of Filipinos to help pay for the debt, despite the loneliness, uncertainty, and humiliation they often suffer, is perhaps the worst effect of debt addiction on government policy. The war in the Middle East dramatized this situation with reports of Filipino migrant workers being raped and killed in the zone of conflict.

Future generations will also suffer because of the environmental degradation accelerated by the debt problem. Only one-fifth of Philippine forests are left, due partly to massive exportation and smuggling of logs and other forest products for the sake of generating scarce foreign exchange. Whole mountains are being torn up and formerly productive rivers and lakes are being destroyed by export-oriented mines and other polluting industries. Only one-fourth of Philippine coral reefs are in good condition, and fisheries production has dropped by half as a result of cyanide, dynamite, and other destructive forms of fishing. Debt-connected poverty has driven fisherfolk to resort to such desperate methods, just as the landless rural poor try to eke out a living by encroaching further and further into forest lands. If trends continue, Philippine forests will completely disappear within this decade—which means ever more disastrous floods, droughts, and landslides.

Although the Philippines has been firmly in the debt trap since the early 1960s when the economy was decontrolled, the peso devalued, and other IMF-imposed policies were instituted, the debt crisis became a full-blown one only during the latter years of the Marcos period (1965-1986). In 1965, the country's debt burden was a mere US$599 million. By 1970, this had risen to US$2.3 billion; by 1975, to US$4.9 billion; by 1983 (the beginning of the end for the Marcos dictatorship as waves of protests followed the assassination of Benigno Aquino, Jr., the most prominent opposition leader), to US$24.1 billion; and by February 1986 (when the Marcos regime collapsed and the Aquino administration took over), to US$26.3 billion.

Two factors contributed to this escalating debt burden. One was the over-eagerness of the big American, Japanese, and European banks to lend billions of petro-dollars to developing countries, including the Philippines, in the 1970s. The other was the borrowing sprees which Marcos' allies engaged in to finance business empires which piled up debts which were later passed on to the government to assume. These two factors aggravated the long-standing problem of a balance of payment deficits caused by trade deficits and profit remittances, which had to be financed by more and more foreign borrowings.

It was during the Marcos period that the government embarked on "development projects" funded by foreign loans. Aside from having an adverse impact on the environment, these projects also harmed and displaced the people in the areas affected. The World Bank, for example, extended a US$61 million loan for the Pantabangan Dam in the central part of Luzon island in 1974. As a result of the project, a whole town was placed under water, and some 14,000 people, who used to earn a decent livelihood from rice, vegetables, and small crafts, were relocated to unsuitable land. Two Mennonite church workers visited the relocation site in 1978, and reported thus:

this World Bank supported project has brought brokenness, the end of a secure village life and no hope for the future. Over 100 of the houses built by the government for the people have slid down from the hillside and been destroyed because the ecological balance developed over the centuries was destroyed. When the hard rains came, nothing was left to hold the loose clay in place.[3]

The World Bank and the Asia Development Bank financed the Agus River Hydroelectric Project completed in 1977. Its aim was to supply power primarily to National Steel Corporation, Kawasaki sintering plant, International Harvester, USIPHIL, and other transnational corporations in the southern island of Mindanao. To assure a stable flow of water essential to power generation, an intake regulation dam had to be built on Lake Lanao, the headwaters of Agus River, which would raise or reduce the surface level of the lake as the need arises. The manipulation of the lake's water levels had a disastrous impact on the Maranao communities who live around the lake. These indigenous peoples, who practice Islam, depend on the lake in many ways:

> Their mosques are built on the water's edge to facilitate the daily ablutions required by their religion; a rise in the level of the lake would flood the mosques, while a fall would separate worshipers from the lake by many meters of muddy flats. The best agricultural land is the basak, the marshy flat land used for wet-rice agriculture, watered by the tributaries feeding into the lake from the mountains. Lowering the lake to a level that affects the water table will diminish the traditional productivity of the land, while raising the level of the lake would obviously have a drastic effect as well on the other source of the region's food, fishing, by destroying the habitat of some fish and permitting the rapid expansion of others. River transportation is also made difficult by shorelines that alternately overflow and recede from banks and docking facilities.[4]

A year after the dam's installation, the lake's level fell by two meters. The Maranao farmers could not plant rice because the land was too dry. The fisherfolk could not catch any fish because the lowered water levels destroyed breeding grounds. Starting in 1978 they had to pay water bills, when before use of the lake's waters for drinking, washing, and bathing had been free. In desperation, many were forced to go elsewhere to find other forms of livelihood.

Another World Bank project, to construct dams on the Chico River, would have destroyed forests and wiped out at least ten Kalinga and six Bontoc settlements.[5] Tribal opposition to the project cost blood and tears. Even while the area was still being surveyed, Kalinga women walked into a surveyors' camp and dismantled their tents. They were arrested and detained together with their men who were also resisting the project. In 1980, the most prominent leader of the resistance was murdered. But the people pushed on and eventually prevailed: the Marcos government and the World Bank eventually canceled the project.

Failure to prevail in the conflict over the Chico dams, however, did not stop the government from embarking on other projects with potential adverse impacts on local populations and the environment. In 1981, it borrowed from British and French commercial banks to finance a dendro-thermal development program. The aim was to generate electricity through wood-burning power plants, which meant cutting down trees for wood fuel.[6]

It was also during the Marcos period that the government adopted export orientation as a development strategy and as a way of acquiring dollars to cope with a worsening debt problem. In the decade of the seventies, wanton destruction of Philippine forests and other natural resources reached its height. From 1970 to 1979, out of US$244.106 billion in total exports, a cumulative average percentage of 81.4 percent or US$18.838 billion was accounted for by resource-based exports, primarily forest products worth US$3.342 billion and mineral exports worth US$4.168 billion.[7] These figures are even understated because of rampant smuggling of logs, particularly to Japan. During the last years of the seventies, and the beginning years of the 1980s, log shipments based on official Philippine statistics comprised only thirty-eight percent to sixty-four percent of their value in Japan.[8] It is not surprising that the amount of forest lands leased out to loggers under timber license agreements (TLAs) increased from forty percent in 1970 to ninety percent in 1977.[9]

Same Problems under Aquino

The Aquino administration, instead of making a clean break with the Marcos past, followed in its footsteps. President Aquino promised to pay the Marcos debts "if only for honor." Even after Congress passed a joint resolution pressing for a thirty-month debt service moratorium to enable the country to recover economically and to rebuild areas devastated by the 16 July earthquake, Aquino and her debt negotiators from the Central Bank and Department of Finance refused to heed Congress's call for fear of adverse reactions from foreign creditors. In general, her administration followed many of the same economic policies closely identified with IMF-World Bank prescriptions as had the previous regime.

The Aquino government continued the Marcos tradition of building environmentally destructive "development projects." The Philippine National Oil Corporation, whose large-scale energy projects rely mainly on external funding, plans to build a geothermal plant on Mt. Apo, the sacred mountain of the Lumads in the southern island of Mindanao. About 400,000 Lumads will be affected by a project which completely ignores their right to their ancestral domain. Mt. Apo, the highest mountain in the Philippines, is also noted for the diversity of its flora and fauna, and has been declared a national park, a sanctuary for the fast-disappearing monkey-eating eagle, and a recognized "environmentally significant area" by the United Nations and ASEAN. The geothermal plant, therefore, threatens not only the Lumads, who have made a protest caravan against it, but also the wealth of species which claim the mountain as their home.[10]

Resource-based exports continued to flow out of the country in the late 1980s under Aquino. Out of total exports of US$57.064 billion during this period, a cumulative average of 61.2 percent, or US$29.441 billion, was accounted for by resource-based exports, principally forest products, wood manufactures and furniture and fixtures (US$4.419 billion), and mineral products (US$4.783 billion).[11] These figures do not take smuggling into account.

Expectedly, wanton destruction of Philippine forests continues unabated. According to Senator Aquilino Pimentel, one of the main proponents of a total logging ban, "twenty-five hectares of forests are destroyed per hour. Some 420,000 hectares have been ravaged in the past two years while only 150,000 hectares have been replanted."[12] Palawan island, the "last remaining environmental frontier of the Philippines," is being denuded of its rainforests by commercial logging at the rate of 19,000 hectares a year. Its rich stores of flora and fauna (232 species), some of which can only be found there, is under threat, prompting a "Save Palawan" campaign mounted by environmental groups.[13]

The debt is a big factor in continuing forest destruction. The government has very limited resources for reforestation because debt servicing eats up a huge portion of the national budget. Secretary Fulgencio Factoran, Jr., of the Department of Environment and Natural Resources (DENR) was even quoted as saying that his ministry does not have the capability to enforce a total log ban. The debt and the dollar scarcity it creates also explains the reluctance to consider the importation of wood for local industry needs while a total log ban is in place.

As far as destructive mining is concerned, the experience of the Cordillera region in northern Philippines is perhaps the most telling. In this region is concentrated some seventy-three percent of Philippine gold production, forty-six percent of silver production, and twenty-two percent of copper production. Transnational corporations and their Filipino partners control the mines whose profits go out of the Cordillera, if not out of the country. The mines prefer to employ lowlanders and suspect indigenous peoples of "highgrading" or stealing ore.

The mines, being "voracious users of pine logs" as timbering for underground tunnels, have destroyed much of the surrounding forests.[14] Logging concessions in the "pine belt" aggravated the situation. Today, about half of the Cordillera region's forests are denuded, and soil erosion from slight to severe affect more than a third of the region.[15]

Large-scale mining has other adverse effects including deformation of the natural landscape as a result of soil erosion, reduction of water supply (with the dropping of the water table), siltation, chemical pollution of rivers, and destruction of plant and aquatic life. The Bued and Agno rivers, to which the mines in the Cordillera empty their wastes, are already dead. Harmful chemicals—such as mercury, lead, cadmium, arsenic, and cyanides—from the mines flow into water bodies which are used for irrigation and domestic purposes. The adverse impact on the health and agricultural produce of the indigenous communities still has to be measured.[16]

The latest and most severe environmental threat comes from open-pit mining, where excavations are made on the earth's surface to expose the gold ore veins. Benguet Corporation's P600 million Grand Antamok Project (GAP) will scrape a large mountainous area flat, scooping up millions of metric tons of earth and dumping it into surrounding areas, divert a major river and channel it elsewhere so that the company can mine its bed, dry up water sources, and destroy the land on which the indigenous Ibaloi and Kankanaey peoples built their small-scale mines and grow their food.

The communities affected mounted a vigorous campaign, culminating in a human barricade to prevent the passage of trucks bearing gold ores. Benguet congressman Samuel Dangwa, however, said members of Congress and other national officials are not inclined to favor GAP's closure because of the dollars it could generate. According to reports, Dangwa made remarks to the effect that "BC's mining activities help bring in the much-needed foreign exchange that can help pay off the country's almost US$30 billion debt."[17] Benguet Corporation itself is reportedly heavily indebted to Bankers Trust, Bank of America, and Export Development Corporation.[18]

Debt, Poverty, and the Environment

When we talk about the poor, we are talking about at least half of the Filipino people. According to official data, poverty incidence as of 1988 was 49.5 percent. Unofficial sources, however, counted seventy percent of the population as living in absolute poverty, meaning that they "cannot buy for their families recommended nutrient requirements, cannot permit two changes of garments, cannot permit grade six schooling for the children, cannot cover minimal costs of medical care and cannot pay for fuel and rent."[19]

How does the debt affect the poor? First, because of the huge amount that the government has to allocate for the debt, very little is left to provide for the needs of the poor. Working on the 1989 budget figures (43.9 percent for debt service versus 38.7 percent for economic and social services), one source claims "each family loses P10,000 worth of government services simply because of the foreign debt."[20]

The poor go to severely underfunded government schools and government hospitals, if they can, because they cannot afford to pay fees for the much better private educational and medical services. In 1989, the Department of Education, Culture and Sports received only P24.675 billion, and the Department of Health, only P7.024 billion compared to P97.712 billion for debt servicing.[21] The problem, however, does not end here. Out of the allocated sums for social services, how much actually goes to the poor (conservatively estimated at fifty percent of the population) after subtracting the salaries of government personnel supposed to deliver these services as well as maintenance and other operating expenses? Only P880 per capita or P5,280 per family.[22]

This figure is both minuscule and unjust, when compared to what the government takes from the poor in the form of taxes (P8,000 per family),[23] much of which is exacted by the government in response to debt-related financial difficulties. The system of taxation is regressive, meaning that indirect taxes on goods and services (of which the poor consume more) account for a greater percentage of total taxes compared to income and property taxes. The poor pay taxes every time they take a jeepney ride, every time they use gas to cook a meal, every time they see a movie, every time they buy a pack of cigarettes or a piece of candy.

The bad news is that more taxes have recently been put in place. "Sin taxes" have been slapped on liquor and cigarettes. These are in compliance with the provisions of a "Letter of Intent" which contains the government's economic policies in response to IMF conditionalities. The LOI seeks to raise revenue collection by P25.3 billion, not only though new taxes, but also by increasing service charges provided by government corporations, which means higher electricity, water, and port rates, which the poor also pay for.

How else are the poor affected by IMF-imposed policies? Removal of government subsidy for and control over the price of rice has already resulted in a steep price hike. Filipino families, who rely on rice to provide the bulk of their meager diet, now face the prospect of increasing hunger. The rapid devaluation of the peso vis-à-vis the dollar means a drastic rise in the prices of imports, primarily oil, industrial raw materials, and equipment. This drives up the cost of transportation and prime commodities. With devaluation and the resultant inflation, the purchasing power of the poor is fast decreasing. Whatever increase in pay or income they manage to win through concerted effort is seldom sufficient even just to regain what they have lost. Workers who are unorganized, unheard, and/or invisible, many of them women and children, have to make do with stagnant earnings while prices escalate.

Out of desperation, many Filipinos, conservatively estimated at one and a half million, now work overseas where they can earn dollar incomes many times the maximum they can get were they to remain at home. With the estimated US$2.5 billion they remit annually, in a very real sense, they are the ones paying for the country's debt.

Among the poorest of the poor are the upland cultivators, marginalized farmers, and landless rural folk who resort to slash-and-burn agriculture in the forest areas. The lack of a substantial and thoroughgoing agrarian reform, which is attributed to a lack not only of political will but also of material resources of a debt-ridden government, has resulted in land concentration in the hands of a few and landlessness on the part of many. Deregulation, as prescribed by the World Bank, means removal of subsidies and support services for small farmers, many of whom are ruined by the high prices of imported farm inputs and the low prices fetched by their products. Protracted drought and recurrent floods add to the misery of the rural poor who, in desperation, are often driven upland or flood into the cities in search of livelihood.

According to Delfin Ganapin, Jr., an official of the Department of Environment and Natural Resources, some seventy percent of the poorest households in the country get their food and fuel from forests.[24] Does it follow that they are primarily responsible for deforestation? One study points out that shifting cultivators settled mostly in residual forest areas already partially cleared by previous commercial logging activity. It answers the question by posing another question: "To what extent are poor migrants, with their simple tools and family labor, capable of cutting the standing timber?"[25]

Another sector frequently mentioned as both victims and culprits are the fisherfolk. Philippine waters can no longer yield as much for small-scale fishing because of rapid environmental degradation. Here, debt figures in two ways. First, the export of forestry, mineral, and fishery resources to earn much-needed dollars has resulted in extensive siltation, pollution, salination, and destruction. Second, the intensifying poverty associated with debt-fueled development has driven marginalized fisherfolk towards using cyanide, dynamite, *muro ami*,[26] and other harmful methods of fishing.

Of the estimated half a million hectares of Philippine mangrove forests, only 146,139 hectares are left as a result of encroachment by fishponds and other forms of aquaculture. Mangrove forests are important because they serve as breeding, feeding, and nursery grounds for fish and other aquatic life.[27] In 1988, the total aquaculture area was estimated to have increased to 224,264 hectares, which means more and more mangrove forests are being converted to fish farms. For every hectare thus converted, a maximum of 1.4 metric tons of fish and shrimp is lost annually in terms of potential production. Worse, fish farms occupy the former fishing grounds of small fisherfolk, who are forced to fish somewhere else in ever-constricting space. They cannot be absorbed in the fish farms, which are capital-intensive, not labor-intensive.[28]

Many of the fish farms cater to foreign markets by engaging in prawn culture, in conjunction with the government's export-oriented fishery policy encouraged by Japanese loans.[29] Lately, prawn farms have been under fire because they cause the destruction of surrounding fields through salination.

Only twenty-five percent of Philippine coral reefs are in good condition, and some fifty percent of them are in "advanced stages of destruction."[30] These reefs serve as home and breeding ground for fish. They are being destroyed by siltation due to soil erosion caused by forest denudation, by the use of cyanide, explosives, and *muro ami* techniques to get quick returns in fishing, by pollution from assorted sources, and by quarrying for construction, aquariums, and export purposes.

The decline in coral reef production by at least thirty-seven percent in the last thirty years has resulted in the loss of seafood that could feed three million Filipinos. Ironically, the fisherfolk who destroy coral reefs and small fish in the course of dynamite fishing (about fifty percent of fisherfolk are said to resort to this), so they can catch more and thereby live through the next day, are at the same time destroying the very base of their sustenance over the long term. Thus, the cycle begins: poverty breeds destruction which breeds more poverty.[31]

Many of the country's lakes and rivers are dying because of siltation resulting from overcut forests and pollution coming from industrial plants, and mine tailings, fertilizers and insecticides used in agriculture, and from domestic wastes due to inadequate garbage disposal and sewage systems. Laguna de Bay, Southeast Asia's largest inland lake, formerly a source of livelihood for some 11,000 fishing households in twenty-seven lakeshore towns, is now dying and can support fewer and fewer fisherfolk.[32] About forty of the country's 400 rivers are "virtually dead."[33]

Towards a Sustainable Development Strategy

In the Philippines, there are two big networks focusing on debt and environment. One is the Freedom from Debt Coalition (FDC), which counts on the support of 144 non-governmental and people's organizations, and on the expertise of prominent academics, sympathetic politicians, highly skilled researchers and economists, and dedicated social development professionals. Its minimum program includes a moratorium on foreign debt service payments, disengagement from fraudulent loans, and limitation of debt service to not more than ten percent of export earnings. Its mission is two-pronged: to "work towards the realization of a people-oriented debt policy through the building of a Filipino freedom from debt constituency, strengthening international linkages and partnerships, and linking the debt policy to a comprehensive alternative development strategy," and "to work with other national broad formations, coalitions and organizations to realize genuine and sustainable development in the country."[34] The other network is the Green Forum-Philippines, a coalition of non-governmental and people's organizations, church groups, academic, and research institutions.

How do the environmentalists see the connection between debt and ecology? In an interview, Maximo Kalaw, Jr., president of Green Forum, explained that "the whole concept of debt service amounting to forty-three percent of the budget is not going to make a society sustainable, especially if half of them are poor as in the Philippines." According to him, the situation is even worse "if the very resources to pay this debt are coming from the destruction of the natural resource base, for two reasons." One, "most of the natural resources in poor countries like the Philippines are very undervalued." In the case of logs, he says, "they do not cost the value of the top soil that is lost together with these, the value of water conservation associated with these, and of carbon dioxide conversion, the loss of marine resources, marine harvests, etc." Secondly, "it is not the same guy who benefits from the resource who pays the cost. In other words, the logger does not pay the cost and in fact, pays very little for the logs. Nature provided the capital so he does not mind destroying it." The ones who shoulder the cost, according to Kalaw, are basically the majority of the people around the resource base who become poorer because of the effects, for example, of forest denudation in terms of the soil that was eroded and the decline of water for agriculture. "So we've got the social inequity component of the repayment portion." In this sense, he says, "foreign debt is really one of the biggest poverty-making machineries there is in Third World countries."[35]

According to Kalaw, the poor who are most affected by environmental degradation comprise the natural constituency of the green movement. Among them are the "four million tribal people affected by the logging policy, eighteen million people squatting in the uplands, and two million fishermen marginalized from coastal fishing grounds." They are in the best position to manage and protect the resources from which they derive their sustenance and livelihood. Thus, Green Forum has advocated "the recognition of tribal land rights and the inclusion of tribal people in the management of protected areas, and an immediate ban on logging of all virgin forests and the transfer of the rights to forest resources to upland communities."[36]

As regards the debt crisis, Green Forum has a number of concrete proposals: "The shifting of debt servicing funds to development through a debt-for-development swap, the shifting of official development assistance (ODA) funds from military and government infrastructure spending to funding community capability building by NGOs, and the conversion of poor country's debts to 'Common Futures' indemnity bonds for an NGO-PO (Non-Governmental Organization-People's Organization) development fund."[37]

Green Forum has already figured in such a swap. World Wildlife Fund-USA bought US$2 million worth of debt which, after being redeemed by the Central Bank in local currency, would be channeled to conservation projects jointly approved by Haribon Foundation Inc. (one of the leading organizations in Green Forum), the Department of Environment and Natural Resources, and the World Wildlife Fund.[38]

Both Green Forum and the Freedom from Debt Coalition are actively forming and strengthening linkages with like-minded and sympathetic groups in both the North and the South. Debt and environment, after all, are global issues which call for a concerted global response. As the Brundtland Report succinctly put it:

> The onus lies with no one group of nations. Developing countries face the obvious life-threatening challenges of desertification, deforestation, and pollution, and endure most of the poverty associated with environmental degradation. The entire human family of nations would suffer from the disappearance of rain forests in the tropics, the loss of plant and animal species, and changes in rainfall patterns...All nations will have a role to play in changing trends, and in righting an international economic system that increases rather than decreases inequality, that increases rather than decreases numbers of poor and hungry.[39]

Notes

1.   Leonor M. Briones, "The Philippine external debt: International cooperation toward a people-centered debt strategy," paper presented to the US-Asia Institute, Philippine International Convention Center, 14 May, 1990.

2.   *Women Want Freedom from Debt—A Primer* (Quezon City: Freedom from Debt Coalition, 1989).

3.   Cheryl Payer, *The World Bank—A Critical Analysis* (New York: Monthly Review Press, 1982), p. 305.

4.   Payer, *The World Bank*, p. 268.

5.   Annex A in Ponciano L. Bennagen, "Philippine cultural minorities: Victims as victors," in Vivencio R. Jose, *Mortgaging the Future—The World Bank and IMF in the Philippines* (Quezon City: Foundation for Nationalist Studies, 1982).

6.   Leonor M. Briones, "The continuing debt crisis and the destruction of the environment: Two aspects of the same coin," unpublished paper, p. 9.

7.   Briones, "The continuing debt crisis," p. 11.

8.   Briones, "The continuing debt crisis," p. 16.

9.   Germelino M. Bautista, "The forestry crisis in the Philippines: Nature, causes and issues," typescript, p. 10.

10.   "Dayandi: To the last drop," in *Tribal Filipinos and Ancestral Domain—Struggle Against Development Aggression* (Quezon City: Tabak, 1990).

11.   Briones, "The continuing debt crisis," p. 11.

12.   "For a balanced ecology, ban logging for 25 years," in *Newsday*, 15 March, 1990.

13.   *Save Palawan*, leaflet, produced by the Haribon Foundation.

14.   Abe P. Belena, "The systematic destruction of the fragile cordillera," in *Sarilakas Grassroots Development*, Vol. 4, Nos. 1/2.

15.   *ITAG Ecowatch*, Vol. 1, No. 1, 1989.

16.   Paul Valentin, "The state of the environment in the Cordillera," typescript, 1987.

17.   "Military crackdown imminent on Itogon folk protesting open-pit mining," *Northern Dispatch*, Vol. 2, No. 12, 1990.

18.   Papers of the founding congress of the Cordillera Committee for Environmental Concerns, March, 1988.

19.   *Ibon Facts and Figures*, Vol. 11, No. 24, 1988.

20.   Karina Constantino David, "Access of the poor to basic social services," in *Poverty and Growth in the Philippines* (Manila: Friedrich Ebert Stiftung, 1989), p. 34.

21.   Tentative figures based on Table II. B. I, National Government Expenditures by Department—Special Purpose Fund 1976-87 (with estimates up to 1990), Department of Budget and Management.

22.   Constantino David, "Access of the poor," p. 36.

23.   Constantino David, "Access of the poor," p. 36.

24.   Briones, "The continuing debt crisis," p. 14.

25.   Bautista, "The forestry crisis," p. 24.

26.   In *muro ami*, deep-sea divers, including children, pound coral reefs with rocks in order to drive fish into waiting nets.

27. Alfredo R. Pascua, "A hard look on our ills and tribulations," in *Sarilakas Grassroots Development*, Vol. 4, Nos. 1/2.

28. "Aquaculture and the subsistence fisherfolk," in *Lundayan*, Vol. 1, No. 1, 1990.

29. The tenth OECF loan consisting of five billion yen was for the Agro-Industrial Technology Transfer Program (AITTP), a large chunk of which (thirty-two percent or almost P170 million as of mid-1989) went into financing prawn farms ("The top-down mechanism of agricultural development aid," in *AMPO Japan-Asia Quarterly Review*, Vol. 21, No. 4, 1989.

30. "Aquaculture and the subsistence fisherfolk."

31. *Coral Reefs*, Fact Sheet of the Haribon Foundation.

32. Aleli Bawagan, "Laguna Lake for whom?," in *Philippine Currents*, October, 1989.

33. Pascua, "A hard look."

34. *Strategic Plan of the Freedom from Debt Coalition Approved in its Second Congress*, July, 1990.

35. Interview with Maximo Kalaw, Jr., Green Forum Office, Makati, Metro Manila, 19 April, 1990.

36. *Creating a Common Future—Philippine NGO Initiatives for Sustainable Development*, document of the Green Forum.

37. *Creating a Common Future*.

38. The Haribon Foundation, *President's Report, February 1988-February 1989*.

39. World Commission on Environment and Development, *Our Common Future* (New York: Oxford University Press: 1987), p. 22.

12

Economic Growth and Environmental Problems in South Korea: The Role of the Government[1]

Dong-Ho Shin

Industrialization was considered to be an essential condition for economic growth in the development literature of the 1950s.[2] Economic growth, it was assumed, would automatically bring a better quality of life. Most state governments strongly committed their efforts to national industrialization in the past several decades. Some countries (particularly the Newly Industrialized Countries in Asia) devoted tremendous energy to growth and were highly successful economically.[3] However, few of them paid attention to other aspects of development, such as environmental quality. As a consequence, these countries have had to struggle with an environmental crisis. South Korea (hereafter, Korea) is a good example of this.[4]

The current chapter examines the role of the Korean national government in economic growth and environmental preservation. In particular, it considers the potential for strong government intervention to protect the environment, in a way reminiscent of state intervention in the economy of the 1960s and 1970s. The following section begins with a review of social and economic transitions during the past three decades. This is followed by an examination of governmental actions which influenced the process, including evolving traditions of economic planning. The third section identifies emerging problems in the physical environment and reviews the rise of public concern for the environment in Korea. The fourth section analyses the government's response to the emerging environmental crisis and to the increasing public demand for a cleaner environment. The last section will end with some remarks on the effectiveness of the governmental approach.[5]

Governmental Approach to Economic Growth in the 1960s and 1970s

Transitions in Social and Economic Conditions

Korea experienced very rapid economic growth during the three decades since the 1960s.[6] The change in the Gross National Product (GNP) is a measure of this economic growth (Figure 12.1). GNP per capita grew from US$81 in 1960 to US$5,569 in 1990 at current prices, while GNP grew by 6.9 percent per annum in real terms. There was phenomenal growth in trade with other economies as well. The value of merchandise exports increased 34.1 percent per annum in the first decade since 1960 and another twenty-two percent in the second decade. The growth in the economy was the result of a period of rapid industrialization. Between 1960 and 1990, the number of employees in the manufacturing sector grew from 290,000 to approximately 3,000,000. This growth in manufacturing has, in turn, led to rapid urbanization. Indeed, those who lived in cities with a population of more than 50,000 increased from 28.3 percent in 1960 to 74.3 in 1990 (Table 12.1). In thirty years Korea has shifted from an agrarian to an urban industrial society.

FIGURE 12.1 Korean GNP Growth Rates (in Real Terms), 1954-1990

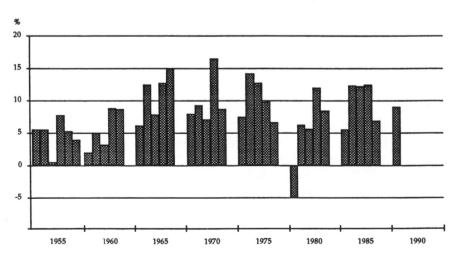

Source: *Korean Statistical Yearbooks,* Various Years.

TABLE 12.1 Selected Indicators of Economic and Social Transformations in Korea

	1960	1970	1980	1990
GNP per capita (US $)	81	242	1,510	5,569
Export (US $ in million)	.03	0.8	17.5	69.8
Import (US $ in million)	.34	2.0	22.3	65.0
Agriculture in GNP (% in total)	35.9	26.2	15.8	10.3
Farming households (% in total)	58.3	44.8	27.3	19.5('89)
Manufacturing employment(% in total)	N.A.	9	28	28.2
Urban population (% in total)	28.3	43.1	57.3	74.3
University students (thousand)	92.9	146.4	402.9	1,040.2
No. of Motor-vehicles (thousand)	31.1	144.3	527.7	3,339.4
Total population (million)	25.0	31.5	37.4	42.8

N.A.: Not Available

Sources: *Korea Economic Report* (Seoul), Vol. 6, No. 6, 1991; United Nations, Department of International Economic and Social Affairs, Statistical Office, *Monthly Bulletin of Statistics*, 1991; Food and Agriculture Organization of the United Nations, *FAO Yearbook: Trade and Commerce*, Vol. 44, 1990; Economic and Social Commission for Asia and the Pacific, *Statistical Yearbook for Asia and the Pacific*, 1990; *Korean Statistical Yearbook*(s), 1961, 1971, 1981, and 1991; Bank of Korea, *Economic Statistics Yearbook*, 1980.

Economic growth in Korea was accompanied by an improvement in social conditions. The number of people per physician decreased by half (3,219 in 1960 to 1,392 in 1990), while life expectancy increased from fifty-four to sixty-six years. During the same period, the adult literacy rate increased from seventy-one percent to ninety-four percent, while students enrolled in four-year universities grew from four percent to fourteen percent of the population aged twenty to twenty-four. These changes in Korean society did not happen by accident. Rather they were outcomes of systematic planning and close monitoring by the Korean central government. A brief review of Korean economic policies illustrates the forces which have driven these social and economic changes.

Korean economic growth in the post-independence era (since 1945) can best be divided into three periods that correspond generally to the nature of the political regimes. The two most critical divisions occurred in 1960 and 1980. In 1960, a leadership generation (e.g., Syungman Rhee) raised in a traditional society gave way to a strong military regime (that of Chung-Hee Park), equipped with modern organizational skills.[7] In 1980, a new generation of leaders, sensitive to changing social and environmental conditions, emerged. An understanding of the twenty year period separating these two events is necessary to grasp current changes in Korean social and economic structures. During most of this period, the government was led by the authoritarian Chung-Hee Park. His regime initiated systematic planning and monitoring with specific goals and targets of development. This approach showed signs of success from its inception. In contrast, the indecisive efforts of earlier governments had contributed to the lackluster performance prior to 1960. Therefore, it is worthwhile to elaborate on Korean economic policies beginning from the early 1960s.

Governmental Role in Economic Growth in the 1960s and 1970s

The Korean economy took off in the early 1960s. This take-off was driven by a strong commitment to economic growth by the political leadership which emerged from the 1961 military coup. Their initial focus was to improve the poor social and economic conditions which had not changed in centuries. The leadership made important changes in the organizational structure of national planning. The most critical one was to reform central planning institutions. The Supreme Council of National Reconstruction, the primary instrument of the 1961 coup, created the Economic Planning Board by merging the existing governmental agency (the Economic Development Council) with ministerial subsidiaries of planning, budgeting, and statistics.

The Economic Planning Board exercised an unusually wide range of authorities: planning the national economy, budgeting finances, and filing national statistics. In 1961, the Economic Planning Board was directed to prepare a series of five-year economic plans. The First Five-Year Economic Development Plan (1962-1966) set a target of 8.3 percent annual GNP growth. In the planned five years, the economic performance exceeded the originally planned target. The Second FYEDP (1967-1971) was formulated in 1966. The new plan projected 7.0 percent annual increase in the GNP and emphasized changing governmental assistance for the industrial sector. A difference from the previous period was made in moving the focus of growth to industrial growth from the primary sector. With the successful implementation of planned programs and projects, the Korean economy repeatedly set records in GNP growth rates.

With some experience from the previous planning periods, the Third FYEDP (1972-76) was formulated in 1971. The basic policy (that of industrialization) was unaltered. The third planning period was eventually the most successful in economic terms. Korea achieved a 10.1 percent annual increase in GNP, far beyond the planned target of 8.6 percent. However, for the first time negative effects of the fast economic growth began to emerge. The major problem was urban pollution, a result of unforeseen urban population growth and planned industrial growth. The preceding developmental process, unfortunately, did not consider this problem, much less attempt to eliminate it. The plan for the fourth period (1977-81) projected a 9.2 percent growth in the GNP, with the principal priority on industrialization.

The regime did not survive the October 1979 assassination of President Chung-Hee Park, the driving force of systematic planning and strong commitment to national economic growth. By then, Chung-Hee Park had been in office for eighteen years consecutively. Throughout the whole period, Park's regime exercised strong central control. His authoritarianism had grown even stronger in the second half of the 1970s in response to increasing criticisms from opposition leaders, university students, and intellectuals. However, no organized agitation towards environmental conservation existed, with the exception of fragmented and localized appeals for compensation associated with environmental pollution. Nor did genuine governmental regulations relating to environmental standards exist at that time. Rather, high ranking governmental officials and economic planners were preoccupied with the notion that environmental regulations hampered continuous economic growth.

During the time between the 1979 presidential assassination and the establishment of a successor regime under Doo-Whan Chun in 1980, a political vacuum existed in Korea. In this period, street demonstrations for greater political freedom and other social demands were expressed. Even at this time, however, no environmental movement emerged. Such a movement only began to emerge slowly during the 1980s.

Governmental Role in Economic Growth in the 1980s

In 1980, when Doo-Whan Chun stepped into the presidential office, economic conditions were unusually poor. From the end of the 1970s, the Korean economy had already started to suffer from skyrocketing inflation, under-use of production capacity from over-investment (especially in the heavy and chemical industry), rising crude oil prices, and increasing foreign debts. The GNP actually declined by 5.7 percent in 1980, the first drop since 1962. In August 1981, the Chun government initiated a number of programs to correct these problems through the Fifth Five-Year Economic and Social Development Plan (FYESDP) for the period of 1982-1986. In this plan, an emphasis was placed on "social" development in addition to solely "economic" growth, although the export-led industrialization was taken as a basic strategy.

Signs of both stabilization and growth followed through the middle of the 1980s. In 1986, Korea had a trade surplus of US$11 billion for the first time in its history. Chun's government formulated a sixth plan (1987-1991) in the same year. This plan projected an average annual growth rate at 7.3 percent, but achieved higher rates in the first two years. In 1987, GNP per capita increased by 12.2 percent. This growth continued through the years of 1988 and 1989 as well, thereby generating a US$14.2 billion trade surplus in 1989.[8]

The new government headed by Tae-Woo Roh stressed qualitative improvement, privatization, and regional equality.[9] However, there were growing barriers to sustained growth. Labor costs rose an estimated fifteen percent to twenty percent in 1987 and another twenty percent to twenty-five percent in the following year,[10] hinting at revived inflation (seven percent in 1988). A balance of trade deficit of US$10 billion was recorded for 1990 and another US$13 billion for 1991. Notwithstanding these negative signs, economic growth in Korea is expected to continue in the near future. The Korea Development Institute, a well-qualified research and policy institution, foresaw that Korea's annual GNP would grow at 6.3 percent to nine percent through the years 1991 to 2001.[11]

Although quantitative growth in the economy continued in the 1980s, it is less significant than qualitative transitions in economic and social conditions. Through the earlier half of Roh's tenure, the role of government had changed significantly. Governmental controls on opposition activities have largely been eliminated. Various restrictions on the labor movement and censorship on the media have also been lifted.[12]

The economic growth under the authoritarian regimes of the past several decades has created another challenge for recent governments. Material affluence made available by economic growth has allowed many Korean people to shift their concern from economic well-being to social well-being. In the 1960s, the common priority of Korean people was simply "earning more money."[13] Since the 1980s, however, this has been changing towards "enjoying life" with a Westernized life style, since most Korean people have already satisfied their basic needs. An important characteristic of these changes is the increased concern for environmental quality. Grass-roots advocacy for preservation of natural areas and stricter controls over air and water quality has spread over the country rapidly in the early 1990s. The following section will discuss the interaction of growing environmental problems with heightened public awareness of these problems.

Changing Public Response to the Emerging Environmental Crisis

From the mid-1970s, some cities in Korea already began to experience environmental problems. This was particularly the case in heavily industrialized cities. A well-known example is the city of Ulsan, a large center of Korean petrochemical and machinery industries since the middle of the 1960s. Discharge of industrial waste water contaminated sea water in coastal areas of the city, affecting fishing activities and aquaculture.[14] These traditional activities were major income sources of people who had not taken part in the industrialization process of the city. Industrial water also polluted the Taewha River which runs through the central area of the city. This water pollution, combined with air pollution caused by heavy industry, seriously affected the productivity of rice farming in the delta of Taewha River.[15] People living in Onsan, developed as a sister city of Ulsan, have reported many incidents of pollution-born skin disease since the 1970s.[16] Similar problems were also found in locations, such as the cities of Pusan, Chinhae, Masan, and Kwangyang.[17]

In the early years of development the public and government officials regarded environmental pollution as an unavoidable by-product of economic growth. They viewed this as the price that Korea had to pay to join the world of developed economies. Most people in Korea fully expected that industrial growth in or around their communities would bring wealth and, in turn, a better standard of living to their communities.[18] In such a social environment, organized activities for environmental protection were discouraged.

Even the people who were directly affected by industrial pollution were unwilling to express their concerns. It was difficult for them to choose between potential economic benefits for all Koreans which could be generated by industrial growth and their personal sacrifice forced by resulting environmental problems. Therefore, many people tended to accept environmental decay for the sake of the national economy. Under these circumstances, government officials and the people in general gave little thought to the possible consequences of environmental degradation which might follow industrial growth.

Furthermore, strict government controls in the 1970s and the early 1980s affected political opposition activities, including even the nascent environmental movement.[19] Such activities gained little attention either from the public, the press, or the government. There were no other institutional systems to act on behalf of those who suffered from environmental pollution. This began to change gradually in the 1980s, and the change accelerated in the early 1990s.[20]

Building a Nuclear Waste Storage

An environmental crisis erupted in November 1991, when the central government decided to build a nuclear waste storage facility on Anmyun Island in the western coastal area. By 1990, Korea had established nine nuclear power plants, which were generating forty-six percent of all electricity generated in the country. It also indicated that meeting the current trends to energy consumption would require fifty-five additional plants by the early 2000s.[21] In 1990, the government found that the existing storage sites were becoming saturated. It decided to construct an additional storage facility on Anmyun Island. The decision was made without any consultation or input from the residents of the island. Although this reflected the conventional decision-making process of the Korean government, this time it met with considerable opposition from local residents.

As the government decision on the waste storage was publicized through the press, some leaders of Anmyun Island residents began to organize protests in November 1990. The movement quickly grew as hundreds of other residents joined the demonstration. The protesters' goal was to force the government to abandon its decision, and one method of protest was the occupation of the town office of the island. As in many other protests, the government did not provide any understandable explanation of the issues. Rather, it simply urged the protesters to stop. This failure to explain further upset the protesters causing violent demonstrations. The central government dispatched thousands of riot police to the island. This led, in turn, to a serious confrontation between the police and the protesters. Approximately 10,000 residents, including leaders of various local clubs, farmers, and students, participated in the protest. In addition to the town office occupation, they set fire to a police station and confined governmental officials.[22]

Eventually, the conflict on Anmyun Island was settled after the government withdrew its earlier decision. Press coverage of the episode was extensive. The coverage greatly publicized not only the low standards of governmental behavior towards the public but also the lack of information on the impacts of nuclear waste. In addition, the event helped increase environmental awareness in general and the treatment of industrial wastes in particular. Such collective action, indeed, has brought a new challenge for the national and municipal governments of Korea in locating sites for the treatment of both household wastes and industrial ones.

Through the Anmyun Island crisis, the government learned that public opposition was becoming a major barrier to implementing energy supply plans. The government found that it had to address residents' opposition. Therefore, it began to search for a better approach to choosing locations of nuclear waste storage and to develop methods of convincing people to accept the decisions. The government also began to educate people that nuclear waste can be less hazardous than the public perceives. Learning about advanced technology for the safe treatment of industrial wastes from other countries became a new task for concerned governmental agencies and the academic community in the field.

Chemical Spill to Piped Water North of Taegu

Another crisis occurred after a chemical substance leaked from the Dusan industrial plant, a part of the Kumi Industrial Complex located approximately forty kilometers north of the city of Taegu. The crisis occurred on 9 April 1991, when approximately thirty tons of waste water contaminated by phenol was accidentally released. The waste water flowed directly into Nakdong River, from which ninety percent of Taegu residents' drinking water is obtained. Neither the company nor the local government gave any warning to the public. Rather, the problem was publicized by reports from those who actually smelled or tasted the chemical either from steamed rice or from household tap water. This chemical spill put the lives of more than 1.5 million Taegu citizens in danger and gained nation-wide attention within a few days. Although the physical impact of the accident was limited to this one city, its impact on public environmental awareness extended to the nation as a whole.

There were several underlying reasons for such wide public attention. First, the lives of a great number of people were threatened by the spill which affected drinking water, a necessity of everyday life. This resulted in people gradually coming to realize the potential magnitude of impacts from a single environmental crisis and the indiscriminate nature of such a crisis. Second, the media thoroughly revealed the process of the company's attempt to hide its inefficient handling of waste treatment. In addition, people were outraged by dishonest government statements and inaction even after the leakage. The media successfully reminded the public to be aware that such problems could be happening or could have happened without public attention. Finally, there were increasing signs of environmental degradation in other cities and even rural areas.

Environmental Degradation by Urbanization

There are additional environmental problems that have posed challenges for the Korean government. Two major causes of the problems are the increased urban population and changing consumption patterns of the Korean people. As shown in Table 12.1, 74.3 percent of Korea's population lived in cities in 1990. In the same year, there were sixty-three cities that had more than 50,000 people. Rapid urbanization caused increasing demands for municipal services. Provision of tapped water, fuel, transportation, and the treatment of household waste are only a few of the issues related to serving the increasing urban population.

Supplying clean, as opposed to sufficient, water has been a continuous challenge for municipal governments for decades. There have been ongoing disputes on the quality of drinking water in urban Korea. While non-governmental organizations and outspoken citizen groups often argued that the quality of drinking water in cities was inadequate, governmental agencies counter-argued that it was adequate for drinking. Since the Taegu crisis, however, these governmental agencies have completely lost their credibility.

Another issue is the management of air quality. The process of burning coal briquettes, the major source of household fuel for decades in cities, aggravated the air quality of urban Korea by producing sulfur-dioxide (SO_2).

Urban air quality has been further challenged by increasing car-ownership. During the 1980s alone, the number of motor vehicles in Korea increased by a factor of 6.4 (see Table 12.1). Although the growth in the number of motor vehicles has been apparent over the whole country, the problem was more serious in large cities. The city of Seoul, for example, had one million registered cars in 1990, while the number for 1980 was only 400,000. This rapid growth in the number of cars has considerably reduced inner city driving speed, thereby generating more automobile emission due to increased idling time. Average driving speed within Seoul was estimated to be nineteen kilometers per hour in 1990, and is projected to be twelve kilometers per hour for 1992 and six kilometers per hour for 1996.[23] These problems make maintaining clean air in Seoul very difficult. As a result, the daily level of sulfur dioxide pollutants, measured by the air quality monitoring signboard installed in front of Seoul city hall, exceeded 0.15[24] parts per million (ppm) eight times during the period between 16 November and 9 December 1989.

Public Perception on the Environmental Quality

The deteriorating urban environmental quality during the 1980s and specific environmental crises in the early 1990s have had their affect on public awareness of the environment. The trend to higher public concern can be tracked through a series of nation-wide surveys administered by the Environmental Administration. The first survey (conducted in 1982) showed that 71.4 percent of the 1,883 interviewees considered that environmental quality in Korea was a "very" or "somewhat" serious problem. Two later surveys conducted in 1986 and in 1988 showed this view hardening. In 1986, and again in 1988, more than eighty percent of those who responded stated that environmental pollution in Korea was a "somewhat," or "very" serious problem.[25] Yet another survey, sponsored by the Ministry of Environment in 1990, indicated that 77.5 percent of the 2,000 interviewees felt that the environmental quality of Korea was a "somewhat," or "very" serious problem.[26] In industrial cities, the percentage is even higher. An area-specific survey, executed in the city of Ulsan, indicated that 93.9 percent of the 443 respondents felt that the pollution problems facing the city were a "somewhat," or "very" serious problem. Furthermore, 37.4 percent of the survey sample indicated that the "environment" should be the most important developmental goal in the future.

A study conducted in the early 1980s showed that most Koreans were supportive of governmental strategies towards rapid economic growth. Two studies administered by the environmental ministry[27] showed that this was true for the period prior to the middle of 1980s, but not since. Those who stated that they would accept polluting industries if the industries were conducive to economic growth in their communities declined from 11.5 percent in 1982 to 9.6 percent in 1990.[28] This tendency is confirmed by the fact that those who would never accept any kinds of polluting industries increased from 31.4 percent in 1982 to 47.4 percent in 1990.[29]

Other studies support the above findings. In a survey conducted by the Institute of Social Sciences at the Seoul National University in 1984, interviewees were asked to indicate how important they felt issues of regional economic growth, urban environmental quality, public transportation, social welfare, and so on to be. Among eighteen issues listed as potential answers,[30] "regional economic growth" emerged as the most important. Indeed, 75.5 percent of the interviewees stated that economic growth was "somewhat" or "very" important, while only twenty-nine percent to 34.7 percent considered environmental issues of urban air, water, and noise as being "somewhat" or "very" important. A survey conducted in 1991, however, revealed that a "clean environment" was the most important issue out of seven social issues listed.[31]

Grass-roots action to protect the environment in Korea is new. Although there have been sporadic activities from organized environmental groups, they received little attention from the media or the public prior to 1990. Since then collective action on environmental issues has grown rapidly. As indicated in an earlier section, this growth of environmental consciousness is largely a result of a series of environmental crises. Growing environmental consciousness is well reflected on the occasions of media coverage on environmental issues (Figure 12.2).

The effectiveness of the growing environmental movement, however, is reduced by those who are too individualistic, opportunistic, and short-sighted. Rather than collectively tackling the environmental problem, many people in Korea try to overcome the problem by simply removing themselves temporarily from it. They tend to cope with it in individualistic manner. For example, in some areas automobile owners are able to travel to obtain good drinking water, while the poor are left to continue drinking polluted water.[32]

FIGURE 12.2 Number of Environmental Incidents Covered by Seventeen Major News Media of Korea, 1980-1990

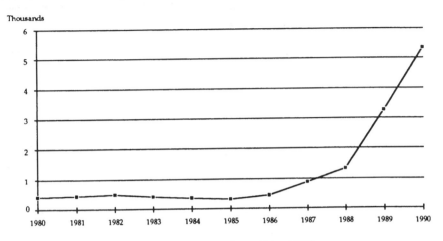

Source: *Environmental White Paper*, Printed in 1991, p. 305.

The increased public awareness of the environment has resulted in better acknowledgment of environmental problems by the Korean government since 1990. Since then, the central government has initiated an examination of the current state of the environment in Korea. There have also been some changes in the organizational structure of Korean environmental policy. This governmental response to growing environmental awareness and its attempts at increasing effectiveness will be examined in the following section.

Governmental Response to Environmental Crisis

Environmental Policies prior to the 1980s

Korea is especially vulnerable to environmental problems because of its high population density. Korea's population density (427 persons per square kilometer) ranks among the highest in the world. However, there has not been strong governmental policies to take this into account. The priority of preserving the environment has always followed that of economic growth. Before 1980, no governmental organizations existed to formulate and implement environmental policy. Fortunately, there has been a gradual change since then. Table 12.2 postulates this change in Korean environmental policy.

TABLE 12.2 Korean Governmental Activities Prior to 1980 and Since 1980

Institution	Scope of Actions	Purpose of Actions	Characteristics of Actions Before 1980	Since 1980
Ministry of Environment	Comprehensive Non-existent	Preventative & Regulating Pollution	Preserving the Environment	
Other Agencies	Fragmented	Regulating Pollution	Weak & Curative	Weak & Curative

Source: Compiled by the author, based on Ministry of Environment (Republic of Korea), *Environmental White Paper*, 1991.

Prior to 1980, Korean environmental policy was fragmented, ad hoc, and sporadic. Environmental policy was narrowly defined as the regulation of industrial pollution, and broader issues of environmental conservation were inadequately addressed. Related activities of environmental agencies were not coordinated at all. Rather, regulating authority was divided among many different ministerial branches. These authorities were further limited not only because of the lack of staff and financial resources, but also because of inter-agency competition. In the 1980s, however, this relatively weak position of environmental management began to improve.

For economic programs, there had been several attempts at forming a governmental agency (or reforming the existing agencies) even before the 1960s.[33] However, for environmental matters, it was not until 1969 that Korea established a small section of environmental health under the Department of Environmental Health within the Ministry of Public Health and Social Work (MPHSW). By 1975, this section was upgraded to the ministerial division level. The Environmental Health Division was further expanded into two departments: one for air pollution prevention and another for water pollution prevention.

In 1977, the status of the Environmental Health Division was further developed through the addition of the Department of Environmental Planning. Although the bureaucracy charged with governing the environment expanded, the MPHSW's activities were still confined to regulating industrial pollution. However, even within this narrow mandate, environmental policies were not strongly enforced because the agency often had to compete with other more powerful ministries, such as construction, or home affairs, or the economic planning board.[34] These ministries took a dim view of environmental regulations.

As a legal body for protecting the environment, Korea first instituted the Law of Pollution Prevention in 1963. This law, however, remained largely unenforced until 1969, when the Implementation Order of the Law of Pollution Prevention was promulgated. In 1977, the 1969 law was replaced by the Environmental Preservation Law, which broadened the scope of environmental policies to embrace preservation of the environment, in addition to the regulation of pollution. In the same year, a new law was enacted to protect the quality of coastal waters. Institutionalizing the Law of Sea-water Protection was necessary since there were increasing reports of contamination of sea-water due to pollutants mainly from industries. In Korea, many industrial facilities were located in coastal areas.[35] The resulting water pollution has been the most serious in areas along the coast of the southeast industrial belt.[36] The 1979 Law of Treatment of Waste Petrochemical Materials was another addition to the existing environmental law system to protect water bodies.

The newly-enacted laws were more comprehensive than the earlier ones, broadening the range of environmental preservation and spelling out specific areas of application. Nonetheless, no sufficient staff or financial resources had been allocated to departments and agencies responsible for exercising these legal authorities (see Table 12.3, for budget allotment). In addition, the institutions responsible for environmental regulations and enforcement remained scattered over several ministries. Those given the task of preserving the environment were, thus, competing on less than favorable terms with better organized ministries. At times they were even in conflict with other priorities within the same ministries! In the late 1970s, therefore, the need for the establishment of an independent agency to solve this complicated circumstance was clear. In 1980, this discussion led to the creation of the Environmental Administration, a governmental agency with a status right below the ministerial level and affiliated with the MPHSW.

Environmental Policies since the 1980s

Since the early 1980s, Korean environment policy has moved to a new phase. The 1980 Fifth Republic declared the people's right "to live in a clean environment" in its amended constitution. Based on this constitutional declaration, the government revised the existing Law of Environmental Preservation to apply to a wider range of environmental issues. The Fifth FYESDP also emphasized this by stating that its goal was "to harmonize economic growth with environmental preservation." It was a fundamental change from the conventional priority given to "economic growth," which was clearly spelled out and strongly supported throughout the previous four planning periods since 1962.

In accordance with the growing governmental concern with environmental issues, there were also some changes in the institutional structure. In 1980, the Environmental Administration created six local monitoring stations and, in the following year, streamlined its operational structure by taking over fifteen previously dispersed environmental agencies. In 1986, all of the existing monitoring stations were upgraded to local branches of the Environmental Administration's implementation structure. The Environmental Administration tightened further its local system through departments and divisions (or sub-divisions) within various levels of local governments.

In 1981, the Environmental Administration developed a research capacity by taking over the National Institute for the Environmental Research[37] from MPHSW. The Environmental Administration also established semi-governmental corporations, such as the Resource Reprocessing Authority, which organizes recycling programs and actually processes used materials.[38] Furthermore, the Environmental Administration drafted more specialized laws and regulations during the 1980s. In 1986, the Environmental Administration prepared the Long-term Comprehensive Plan for the Environmental Preservation (1987-2001), based on some nation-wide studies that it sponsored. In 1990, the Environmental Administration obtained ministerial status with a new name, the Ministry of Environment. In response to the two crises which occurred in the early 1990s, the Ministry of Environment developed a mid-term plan (1991-1996), as an operational element of the long range plan.

Since the early 1980s, Korean governmental activities have shown some success in improving the quality of environment. The government identified polluted places which require intensive management and designated them as "concerned areas." The two largest river systems in Korea, the Han and Nakdong rivers, and some coastal areas adjacent to industrial sites, for example, have been selected for special management. During the mid-1980s, rejuvenation projects been carried out along the Han River, which flows through Seoul and is the major source of drinking water for Seoul's twelve million inhabitants. To preserve the river system, more strict regulations have been enforced on industrial and agricultural activities along the upstream sections of the Han River.[39] Special Preservation Areas for Drinking Water Sources, in which certain activities are banned, were also designated. These efforts have achieved some success, in that the level of water contamination in the Han River declined during the 1980s, while levels of water contamination in other river systems rose (Figure 12.3). The Korean environmental ministry has made progress toward achieving its goal of improving urban air quality as well. The ministry has been persistently encouraging urban residents to replace coal with LNG (Liquid Natural Gas) as the main household fuel source. The former has been one of the three major sources of air pollution in urban Korea.[40] The ministry is also encouraging industry to install waste treatment facilities. It assists industry by arranging tariff reduction rates in importing green technology. To reduce harmful automobile emissions, it encourages oil companies to lower sulfur levels in gasoline. Since the early 1980s, these activities have turned out to be somewhat effective in lowering the level of urban air pollution (Figure 12.4).

FIGURE 12.3 Biochemical Oxygen Demand (BOD) in Four Major River Systems in Korea, 1981-1990

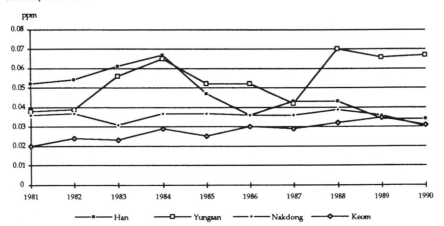

Source: *Environmental White Paper*, Printed in 1991, p. 159.

FIGURE 12.4 Sulfur-Dioxide Contamination of the Air in Five Major Cities in Korea, 1980-1990

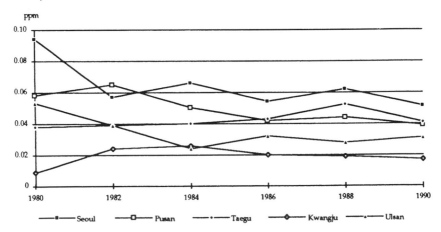

Source: *Environmental White Paper*, Printed in 1991, p. 99.

The Environmental Administration's establishment and its promotion to ministerial status indicate the Korean government's increasing recognition of the importance of environmental policies. This enhanced acknowledgment is also identifiable from the growing amount and proportion of the national budget allocated to environmental policies. As Table 12.3 shows, the environmental budget both in its absolute amount and as a ratio out of the total budget has been continuously increasing at least throughout the 1970s and until the late 1980s. In the last recent two years, however, the proportion of environmental budget has been declining.

The Korean government's gradual recognition on the environment and the strengthened Ministry of Environment's status in the recent years, however, may not be sufficient to correct the environmental problems in Korea. Some environmental groups have criticized the ministry as being inadequately motivated to genuinely protect the environment. Although it has erected standards and regulations, it does not strictly enforce them.

TABLE 12.3 National Budget Allocated to the Environmental Agency, 1971-1990

Year	Amount (Won)	% in total
1971	0.5	0.009
1972	0.9	0.012
1973	0.9	0.014
1974	1.1	0.011
1975	10.6	0.067
1976	13.3	0.059
1977	22.1	0.077
1978	25.1	0.071
1979	51.7	0.100
1980	120.5	0.186
1981	152.2	0.189
1982	207.7	0.216
1983	206.9	0.199
1984	343.1	0.313
1985	420.5	0.343
1986	433.0	0.314
1987	670.8	0.418
1988	772.9	0.420
1989	644.9	0.335
1990	902.1	0.329

Source: Ministry of Environment (Republic of Korea), *Environmental White Paper*, 1991, p. 38.

Some reports criticized that the ministry used the standards and regulations as "window-dressing."[41] Even where the ministry really attempted to enforce regulations, it is often frustrated by its weak presence in certain cases. The neglect of the required Environmental Impact Assessment in the construction of a highway along the western coast, and in the Subway Number Five in Seoul, show that the Ministry of Construction and the city of Seoul, respectively, can ignore the Ministry of Environment's mandate.[42]

Conclusion

The Korean government has promoted ambitious economic plans, and, unlike some other developing countries, Korea has shown an impressive performance. The government since the 1960s rigorously guided the path of development and also exercised authoritarian leadership to prevent social protest (which was believed to slow the pace of economic growth). As the conventional development literature states, industrialization was the engine of growth for Korea. Incidentally, the growth in the economy contributed also to improvement of other aspects of social development, such as education and health care.

Korean economic growth, however, also precipitated a sharp conflict with the environment. Pollution problems were largely underestimated in the earlier planning process. Until very recent years, neither the government nor the public seriously acknowledged these environmental problems. Rather, the government supported industrial growth at the cost of a clean environment. This was not seriously criticized by the Korean people either. From the late 1980s, however, the Korean people finally came to acknowledge environmental problems and began to express their concerns through collective actions. As indicated earlier, these public concerns with environmental degradation were expressed clearly in connection with the governmental decision to build a nuclear waste storage facility on Anmyun Island and a large scale chemical spill into the drinking water of the city of Taegu.

How has the Korean government responded to emerging environment problems? Has its response been as effective as in the creation of economic growth in the 1960s and 1970s? The Korean government since the early 1980s has instituted legal and institutional systems of environmental policy. Through these systems, new programs and projects have been developed for monitoring, regulating, studying, and planning environmental preservation. The Environmental Administration and the Ministry of Environment have played important roles in this regard.

As the Economic Planning Board played a critical role in economic policies since the 1960s, what role will the Ministry of Environment have for the environment? Although the Ministry of Environment has grown considerably since its inception in 1980, its influence is limited by two major factors: the priority of the economy in the national political arena, and the central government's low level of commitment toward the environment. Unless there is a radical shift in the Korean political leadership, economic growth is likely to remain as the important policy objective of the Korean government. Declining proportions of the 1989 and 1990 national budgets allocated to environmental agencies mirror the government's low priority of environmental concerns (see Table 12.3).

Since 1988 President Roh's democratization policies have considerably reduced the influence of the central government. In 1991, local councils with elected members were established at all levels of local government (municipal and provincial) and elections for local governors are also scheduled for 1993. As a result of these changes, future local governments do not expect to be controlled by the central ministries as much as they were in the past. Therefore, the improvement in the quality of the environment will depend more on local governments rather than the central government. This would also delay actions towards environmental protection since it will take a long time for local governments to develop their own institutions and skills to a level that the present central government has achieved.[43]

In this context, the only promising factor which may contribute to environmental protection is the growing grass-roots movements. As the movements in the early 1990s have shown, democratization in Roh's era has already allowed the Korean people to directly and collectively express their demands. Under this circumstance, the people must act collectively, not individualistically. Public action is necessary to make the central and local governments place higher priorities on the environment. They have to show environmental concerns through their voting behavior as well, thereby electing politicians who care for the environment. In addition to influencing government, they have to organize themselves to monitor governmental policies, examine environmental quality, promote an environmental movement, and actually form environmental projects, such as recycling. An active grass-root or people's environmental movement will strengthen and legitimize the ME's claim for prioritizing environmental regulations. If the influence of environmental agencies both at the national and the local government advances, these agencies will be able to mobilize organizational capacity and develop environmental technology, as the EPB and other economic institutions did for the economy.

Notes

1. The author would like to thank Professor Walter G. Hardwick, Dr. P. Devereaux Jennings, Mr Rene Ragetli and Ralph Perkins for their comments on the earlier version of this chapter, and Dr. Sung-Chul Kang for his generous help in collecting data.

2. The most frequently cited work in this area is Walt W. Rostow, *The Stages of Economic Growth* (Cambridge: Cambridge University Press, 1960).

3. See Fredric C. Deyo, ed., *The Political Economy of the New Asian Industrialism* (Ithaka: Cornell University Press, 1987); Alice H. Amsden, *Asia's Next Giant: South Korea and Late Industrialization* (New York: Oxford University Press, 1989); and Brian Heppel, "Pacific sunrise: East Asia's newly industrializing countries," in *Geography Review*, Vol. 3, No. 4, 1990, pp. 7-11.

4. Writings on Korean environmental problems have rarely been available to English readers. The few which are available are Ho Keun Hong, "Who benefits from industrial restructuring?: Reflections on the South Korean experience in the 1980s," in *Korea Journal*, Vol. 31, 1991, pp. 69-84; Seoung-Yong Hong, "Assessment of coastal zone issues in the Republic of Korea," in *Coastal Management*, Vol. 19 No. 4, 1991, pp. 391-415; and H. Jeffrey Leonard and David Morell, "Emergence of environmental concern in developing countries: A political perspective," in *Stanford Journal of International Law*, Vol. 17, No. 2, pp. 281-313.

5. The data for the current chapter was drawn from secondary sources, primarily written in the Korean language. These include newspapers, governmental publications, academic writings, and magazine articles. Survey data collected for earlier studies by other academics has been re-examined by the author.

6. This is well illustrated in Alice H. Amsden, *Asia's Next Giant: South Korea and Late Industrialization* (New York: Oxford University Press, 1989).

7. This has been best described by Hahn Been Lee, *Korea: Time, Change and Administration* (Honolulu: East West Center Press, 1968).

8. Susan A. MacManus, "The three "e's" of economic development...and the hardest is equity: Thirty years of economic development planning in the Republic of Korea (I) and (II)," in *Korea Journal*, Vol. 30, Nos. 8 and 9, 1990. pp. 4-17 and pp. 13-25.

9. Susan A. MacManus, "The three "e's" of economic development," p. 13.

10. *Far Eastern Economic Review*, 19 November, 1987, p. 2.

11. *Comprehensive Mid-term Plan for Environmental Improvement, 1992-1996* (Korean) (Seoul: Ministry of Environment, Republic of Korea, Unspecified Date, printed in 1991), p. 15.

12. This has been widely accepted in the literature on the recent changes in the Korean politics. See Sung-Joo Han, "South Korea in 1987: The politics of democratization," in *Asian Survey*, Vol. 28, No. 1, 1988, pp. 52-61; Sung-Joo Han, "South Korea in 1988: A revolution in the making," in *Asian Survey*, Vol. 29, No. 1, 1989, pp. 29-38; Young Whan Kihl, "South Korea in 1989: Slow progress toward democracy," in *Asian Survey*, Vol. 30, No. 1, 1990, pp. 67-74; and Young Whan Kihl, "South Korea in 1990: Diplomatic activism and a partisan quagmire," in *Asian Survey*, Vol. 31, No. 1, 1991, pp. 64-70.

13. Dong-Geon Byun, "The Korean people's perceptions and beliefs about the environment: Differences among social strata and by sex, community size & educational level," in *Law and Politics* (A Journal Published by the Institute of Law and Social Sciences, Kukmin University, Korea), Vol. 8, 1985, pp. 355-398.

14. The Korean Association of Pollution Problem Studies, *The Korean Map of Environmental Pollution* (Korean) (Seoul: Ilwol Seokak, 1986).

15. The Korean Association of Pollution Problem Studies, *The Korean Map of Environmental Pollution.*

16. Ho-Kyung Jeong; Ji-Ha Kim, *et al.*, *Life or Death* (Korean) (Seoul: Hyungsungsa, 1985).

17. Hong, "Assessment of coastal zone."

18. Byun, "The Korean people's perceptions and beliefs about the environment."

19. Jeong; Kim, *et al.*, *Life or Death*; and Mokpo Association of Green Studies, *To Keep My Earth: Successful Case of Anti-Pollution Movement in Mokpo* (Korean) (Seoul: Pulbit Press, 1988).

20. David A. Smith and Su-Hoon Lee, "Moving toward democracy?: South Korean political change in the 1980s," in Michael Peter Smith, ed., *Breaking Chains: Social Movements and Collective Action* (New Brunswick, N.J.: Transaction Books, 1990), pp. 164-187.

21. Hong, "Assessment of coastal zone," p. 403.

22. Several of the protesters were awaiting trial for these offenses. See *Sae-Kae Ilbo*, 10 November, 1990.

23. *Korea Newsreview*, 17 March, 1990.

24. The maximum allowable standard under the Korean regulation is one third of this amount.

25. Tai-Hwan Kwon, "Perceptions of the quality of life and social conflicts," in *Korea Journal*, Vol. 29, No 9, 1989, p. 13 and pp. 16-17.

26. Daeryuk Research Institute, *People's Perception of Environmental Preservation* (Korean, A survey research sponsored by the Ministry of Environment, Republic of Korea) (Seoul, 1990).

27. Environmental Administration (Republic of Korea), *Comprehensive Long-term Plan for Environmental Preservation, 1987-2001* (Korean) (Seoul) Unspecified Date, printed in 1986; Environmental Administration (Republic of Korea), in corporation with the (Korean) Association of Environmental Education, *A Report on People's Perception of Environmental Preservation: People's Understanding on Environmental Problems Is High* (Korean) (Seoul, 1983); and see Daeryuk Research Institute, *People's Perception of Environmental Preservation.*

28. Environmental Administration, *Comprehensive Long-term Plan for Environmental Preservation*; Environmental Administration, *A Report on People's*, p. 140, for 1983 and for 1990; and Daeryuk Research Institute, *People's Perception of Environmental Preservation.*

29. Environmental Administration, *A Report on People's*, p. 140; Daeryuk Research Institute. *People's Perception of Environmental Preservation*, p. 72.

30. The issues listed include education, crime, and so forth. See the full list in Kyung-Dong Kim, Chung Si Ahn, *et al.*, *Local Development and Local Political Autonomy* (Korean) (Seoul: Seoul National University Press, 1984), pp. 375-376.

31. A survey data set was created by Dong-Ho Shin from the city of Ulsan, Korea, as a part of Ph.D. dissertation research. This includes issues of income disparities, poverty problems, neighborhood safety, and housing. A survey questionnaire is available from the author.

32. Tae-Jun Kwon, "Environmental conscience without collective conscience is nothing," (Korean), in *Sindongah*, March 1990, pp. 386-405.

33. Jong Won Kim, "Perspective for economic development and planning in South Korea," in Andrew C. Nahm, ed., *Studies in the Developmental Aspects of Korea: Proceedings of the Conference on Korea, Held at Western Michigan University, April 6-7, 1967*, (East Lansing School of Graduate Studies and Institute of International and Area Studies, Western Michigan University, 1969).

34. Ministries concerned with the national economy, such as that of commerce and industry, and of construction, have stronger influence on the decision-making process of the Korean central government. See Leroy P. Jones and Il Sakong, *Government, Business and Entrepreneurship in Economic Development; The Korean Case* (Cambridge, MA: Council on East Asian Studies, Harvard University Press, 1980), pp. 58-77.

35. Hong, "Assessment of Coastal Zone."

36. Edwin S. Mills and Byung-Nak Song, *Urbanization and Urban Problems in Korea* (Cambridge, MA: Council on East Asian Studies, Harvard University Press, 1979).

37. In January 1990, this institute had 185 employees.

38. Doo-Ho Rhee, Yun-Hwa Ko, and Luis F. Diaz, "Waste management in Korea," in *BioCycle*, Vol. 27, No. 10, 1986, pp. 44-48.

39. Ministry of Environment (Republic of Korea), *Environmental White Paper* (Korean) (Seoul, Unspecified Date, printed in 1991).

40. See Yoon Shin Kim, "Air Pollution in the Republic of Korea, in *Journal of Air Pollution Control Association*, Vol. 34, No. 8, 1984, pp. 841-843; and Yoon Shin Kim, "Measurements of indoor and personal exposures to nitrogen-dioxide in Korea," in *Environment International*, Vol. 12 1986, pp. 401-406.

41. Mokpo Association of Green Studies, *To Keep My Earth: Successful Case of Anti-Pollution Movement in Mokpo*; and also see *Dong-Ah Ilbo*, 19 January, 1992.

42. *Sisa Jeoneol*, 11 April, 1991.

43. For instance, see the case of the city of Pusan in Hak-Roh Kim, Sung-Chul Kang, and Chang-Won Kim, "A study on the environmental management systems of Pusan, Korea" (Korean), in *Locality and Public Administration Research* (Institute of Local Administration, the Graduate School of Public Administration, Pusan National University). Vol. 4, No. 1, 1992, pp. 81-142.

13

A Social Analysis of the Environmental Problems in Thailand

Suntaree Komin

Environmental problems are, in general, inseparable from the problems of development. Such is the case of Thailand, which has been striving hard to industrialize in the last decade. To the envy of most developed and developing countries, Thailand's economic performance in the past few years with an annual average of a two digit growth rate has been extremely successful. The successful economic growth is mainly due to the explosive development of manufacturing industries, tourism, and direct foreign investment since 1985. However, the result of a research project on the social dimensions of industrialization in Thailand has revealed that the haphazard and opportunistic industrialization experience has had not only many adverse social and economic effects, but has also had dramatic environmental costs, many of which are unfortunately irreversible.[1] With the active encouragement of the government for growth of industries, direct foreign investment, and tourism, the environmental problems have been seriously aggravated on various fronts—from all sorts of pollution from industrial wastes to widespread deforestation and the encroachment of national parks and wildlife.

What happened to the government machinery with regards to environmental controls? Knowing the extent to which the failure to safeguard the environment is due to the negligence of those in the government and private sector industrialists or to their vested interests is essential because of their influence in shaping the development of the country. It is therefore the purpose of this chapter to study the level of environmental consciousness and responsibility of representative government officials and industrialists as well as the educated public in terms of their level of consciousness and cognizance of the causal factors.

Background

Thailand is predominantly an agriculture country, with seventy percent of the labor force still in the agricultural sector. By the early 1990s it had a population of over fifty-seven million, with increases averaging one million per year. The educational background of population at eleven years and over shows that there is a constant seventy-five percent of the population with only primary education, and the gross secondary school enrollment has been a constant low of thirty percent, the lowest in the ASEAN countries. With industrialization, the growth of industrial factories has been concentrated in Bangkok and the five adjacent provinces—the Bangkok Metropolitan Region—compounding population, traffic, and environmental problems.

Environmental Problems in Thailand

Numerous seminars in Thailand on the environment have concluded that the country's environmental problems are due to rapid industrial development and the exploitation of natural resources for export, leading to deforestation, depletion of mineral resources, and pollution of waterways and the atmosphere. The major source of dangerous environmental problems stems from Thailand's estimated 100,000 factories, the majority of which are concentrated in Bangkok and the adjacent provinces. Wastes discharged from these factories in the form of air pollutants, waste-water, and chemical use which directly invades the bodies of people take their toll in short-term and long-term effects. What is the state of environmental problems in Thailand? A cursory look at the following few examples drawn from local newspapers is illustrative.

> A report from a Siriraj Hospital doctor revealed that blood tests on thirty-seven new-born babies during a period of one month showed they had a lead content of ten to twenty-five microgrammes in their bodies. The lead content was reportedly passed on to the babies by their mothers who absorbed it by breathing polluted air and consuming contaminated food and drinks. The acceptable amount of lead in the human body in Thailand is set at thirty-five microgrammes, while in industrial countries like the United States, a lead content above twenty-five microgrammes is regarded as hazardous to health.[2]

A National Environment Board study revealed that ponds near a lead-ore processing plant in Kanchanaburi Province, carried from 200 to 7,000 times the maximum safety level of lead. The untreated water from the ponds flows to Sri Nakharin Dam. A measure below the dam down-stream in the Kwae Yai river was more than four times the lead safety limit of .05 miligrammes per litre. Lead accumulates in the food chain, contaminating fish and other animals.[3]

A two-year-old girl in a flour factory died as a result of inhaling the gas which was produced from aluminium phosphorus used to prevent insects from destroying flour.[4] This is the same type of gas used by the Nazis to kill Jews in World War II.

Orange trees and oranges from thousands of rais of fruit orchards in Pathum Thani Province were destroyed by untreated industrial waste water discarded into the canals by industrial factories around the area. Besides dead trees and rotten fruit, about a thousand farmers who used water from the canal developed rashes, infested eyes, and diarrhea.[5]

Two hundred farmers and residents living near a dyeing factory marched to the provincial hall n Nakhon Pathom Province, where they called on government officials to stop the factory from discharging untreated water into the river, thereby polluting the water needed for their farms and household use. The factory had been discharging pollutants for the last few years and the situation was getting worse. Contaminated water had flooded the whole area, killing fruit trees and causing people to become ill.[6] When the situation remained unchanged, 400 farmers rallied in front of the factory seven months later, resulting in a confrontation.[7]

There is ample evidence indicating the serious rate of deterioration of the environment. The following sections will only briefly review the situation. Detailed studies and figures of the state of environments can be found in various reports.[8]

Water Pollution

Reports of sickness, skin disease, and tainted crops resulting from contaminated water near large factories and industrial zones are widespread in Thailand. Indiscriminate dumping of industrial wastes and chemicals by riverside factories has affected villagers' livelihood and health. Many factories do not install the necessary waste treatment equipment. Where such equipment is installed it is often turned off to save costs. The result is that Thailand's rivers are contaminated with all kinds of filth and garbage, industrial wastes, chemical, plastics, human waste, and disease-contaminated garbage from hospitals.[9] It is no wonder that another National

Environment Board study disclosed that Thailand's rivers are contaminated with coliform, a bacteria that feeds on human waste and which can contribute to the spread of cholera, diarrhea, and other diseases. The level of coliform is now ten times the acceptable level in the Chao Phya river, and as much as forty times more than the acceptable level off Hat Yai in southern Thailand.

The dissolved oxygen level (DO) is the best single indicator of the state of the river's health. The DO level in the Chao Phya river started to deteriorate noticeably some twenty years ago and in the last few years is has been rapidly getting worse.[10] In certain critical areas, like the area towards Bangkok's Port Authority, the oxygen level is almost zero. And the affected areas are continually spreading far and wide. In the Nakhon Pathom area the oxygen level is lower than two milligrams per litre throughout the year. Even Klong Mae Kha in Chiang Mai has also become seriously polluted.

Diarrhea, dysentery, and food poisoning—Thailand's top three diseases—have sharply increased in the last few years. Much of this increase can be accounted for by water pollution (Table 13.1). Besides bacteria pollution of the water, toxicology experts warn that more attention should be given to chemical pollution of water since chemical pollution in water is much harder to deal with or to treat. Pollution by the pesticide paraquat, for example, has resulted in the death of fish in fish ponds over the past few years.

Air Pollution

Air pollution is presently most serious, to the point of being critical, for people living in Bangkok. Over the past few years the air quality has been rated "dangerous" throughout much of the time, with the annual average air quality index nearing 300—the maximum acceptable rate being 100.

TABLE 13.1 Rate of Morbidity of the Top Six Diseases, Thailand (per 100,000 population)

	1981	1983	1985	1986
Diarrhea	513.19	701.58	858.23	1,027.44
Dysentery	91.76	122.01	126.14	155.74
Food Poisoning	45.76	69.84	76.97	85.34
Measles	50.50	69.83	62.21	37.12
Typhoid	23.85	28.26	29.84	13.30
Infection Hepatitis	19.90	22.86	34.56	40.92

Source: Ministry of Health, NESDB, Thailand, 1987.

The major air pollution sources in Thailand are transportation, fuel combustion from stationary sources, industrial processes, and solid wastes. Large quantities of toxic gases and particulate matter are discharged into the air daily by motor-vehicles and factories. Six air pollutants are measured to indicate air quality: carbon monoxide, ozone, sulphur dioxide, suspended particulates, nitrogen dioxide, and lead. Reports from the National Environment Board show that Bangkok, Samut Prakan, Chiangmai, Chonburi, Khon Kaen, and Hat Yai, have pollutants in the air higher than standard. Bangkok's air is polluted with smoke churned out by hundreds of industries and the lead-contaminated exhaust fumes emitted from over one million cars and some 660,000 motorcycles caught in the endless lines of congested traffic. As bad as the situation is at present, the prospects for the future are that it will get even worse. While dust causes irritations and allergies, carbon monoxide can build up to a harmful level and cause headaches, impairment of mental function, impairment of fetal development, and death in the case of very high concentrations.

Air quality is often extremely poor in Thailand's factories. Except in a few modern factories, working conditions in most of Thailand's factories are not satisfactory.[11] There have been countless reports of unhealthy working environments in factories in recent years. For example, a dozen women workers from a factory producing leather goods in the Pathum Thani area were rushed to the hospital in 1989 after they collapsed because they could not breathe. It was found out later that workers in this factory worked fifteen hours a day without break and that there was not enough ventilation or electric fans.[12]

Chemical Pollution

Chemical pollution is another serious threat to people's health, to factory workers in particular. There is an increasing use of chemicals in industry, agriculture, and household products, and those chemical pollutants can pose significant risks to people's health when they reach high enough concentrations. Recent studies have shown that some 26.96 percent of agricultural products and seventy-three percent of marine products were found to be contaminated with toxic substances, mainly Dieldrin and DDT insecticides. Not only does contact with or consumption of such chemicals affect people's health, but they also affect the environment as their concentration builds up over time. There is also the impact of the industrial gases known as chlorofluorocarbons (used as propellants in aerosol cans) which contribute to the depletion of the earth's ozone layer. A National Environment Board study released in 1987 indicated that industrial toxic wastes discharged into the environment would increase one-hundred percent between 1985 and 1990.[13]

While pesticides help increase food production, they also pose hazards to people's health and the environment. One series of tests for pesticide residue in vegetables found that forty percent to ninety percent of the samples tested contained detectable levels of pesticide.[14] All pesticides are potentially dangerous. Insecticides, however, are generally considered to cause more serious and widespread risks because of the quantity used, their acute and chronic toxicity to living organisms, their cumulative properties, and their tendency to remain in the environment for a long period.

Chemical health threats to the factory workers are even more alarming. A survey of twenty-seven chemical factories showed that twenty of these factories used chemicals classified as major hazards according to the World Bank's "List of Hazardous Substances Requiring a Major Hazard Assessment."[15] In most of these factories, emergency and contingency plans for hazard control were totally absent. It is therefore not surprising to find a fatality rate of 53.7 deaths per 100,000 employees in these chemical factories. This is much higher than the overall rate of 33.0 deaths per 100,000 employees per year for all industries in Thailand. A comparison with Britain and the Netherlands shows that the rate of accidental death in the chemical industry in Thailand is twenty times higher (Table 13.2).

There is also the threat of cancer due to long-term accumulation among factory workers—a threat that over time will also lead to the deaths of factory workers. Studies of heath risks in industries indicated that workers who have worked for over five years in chemical factories showed high incidents of eye, lung, and respiratory infections, which could possibly become cancerous in years to come because of the high correlation between exposure to such chemicals and cancer.[16] Given the fact that the number of chemical industries is increasing in Thailand and that the hazards from chemicals are not restricted to factories but also affect those living nearby, this health threat from chemical industries deserves serious attention.

Manganese poisoning provides an example of one threat posed by the chemical industry. In 1964, forty-one workers in a dry-cell manufacturing plant in Thonburi fell ill from inhaling manganese in the factory. By 1984, they were all reported dead.[17] There is no reliable record of how many more workers have been subjected to such chemical poisoning. Manganese poisoning can be measured in blood, urine, stools, and hair. The average manganese level in a person's blood is 8mg per 100ml, and 25mg per liter for urine. A 1983 examination of blood and urine samples of workers in dry-cell manufacturing plants showed that many workers had manganese levels in their blood and urine that were above the normal (Table 13.3).[18]

TABLE 13.2 Risk of Accidental Death in Britain, Netherlands, and Thailand (Number of Fatalities per 100,000 per year)

Activities	Britain	Netherlands	Thailand
Chemical Industry	1.6 (8.7)[1]	2.4 (12.5)[1]	53.7
Building/Construction	9.9	13.1	109.7
At Work	2.1	4.6	31.7[2]
Motor Vehicle Accidents	13.0	16.5	14.2[3]

1. Figures in parentheses include fatalities from the Flixborough and Beek disasters respectively.
2. 1984 figure from Workmen's Compensation Fund.
3. Bangkok figure.

Source: International Labour Organisation, NICE Technical Report No. 8, 1986.

TABLE 13.3 Manganese Concentrations in the Blood and Urine of Workers in Three Dry-cell Plants, Samut Prakan

(mg/100ml)	Blood		(mg/l)	Urine	
	No. Samples	No.> Normal		No. Samples	No.> Normal
1.82-4.55	38	0	9.06-32.51	38	10
0.82-9.21	6	0	9.08-32.51	6	2
4.54-7.31	81	2	4.08-62.44	81	22

Notes: Normal Mn in blood is 8 mg/100ml; normal Mn in urine is 25 mg/l.

Source: Department of Health, Thailand, 1983.

Another common problem is that of lead poisoning. Lead poisoning has always been one of the most important occupational diseases. Poisoning usually results from lead entering the respiratory system and then into the blood stream. High levels of lead in the body can have adverse neurophysiological effects. Studies have shown levels of lead in the blood and urine of a high number of workers that exceed the normal standard level (Table 13.4). Such studies have also indicated high levels of lead in the air in factories.

TABLE 13.4 Concentration of Lead in Blood and Urine of Workers at Five Lead
Acid Battery Plants, Samut Prakan

	Blood			Urine	
(mg/100ml)	No. Samples	No. Normal	(mg/l)	No. Samples	No.> Normal
12.43-142.37	26	1	74.58-487.01	26	25
35.71-90.50	33	1	39.46-316.59	33	22
31.75-108.52	23	6	59.14-422.42	23	17
15.78-99.55	49	1	39.46-293.72	49	20
19.73-55.16	61	0			

Notes: Normal Pb in blood is 80 mg/100ml; normal Pb in urine is 150 mg/l.

Source: Department of Health, Thailand, 1983.

Situations like those discussed above exist because employers either are
ignorant or they neglect the safety and health hazards in the workplace.
The problem is compounded by the generally low educational level of the
workers, mostly drawn from the rural poor, who are often unaware of the
dangers and risks involved with certain chemicals or processes.

Land Use and Deforestation

In the last decade, and particularly in the last few years, Thailand has
suffered considerable loss of fertile cropland and forest as industrialists
and developers have turned such land into industrial sites, resorts, golf
courses, and condominiums. Frenzied land speculation has driven many
poor farmers into landlessness while Thailand's forest areas have shrunk
by almost half since 1961, from 53.33 percent of the total land area in 1961
to 28.03 percent in 1988. The level of degradation as a result of deforestation
was only brought to public attention by the tragic November 1988 mudslide
and flood in southern Thailand in which a heavy rain brought down
thousands of legal and illegal logs from the forest, killing 300 villagers and
destroying a number of villages overnight in several provinces. That
incident prompted the government to stop the high rate of forest destruction
by enacting an emergency decree banning all logging concessions. But, as
always, policy is one thing, practice is a totally different matter. Despite
the logging ban, encroachment of forest, even national parks, continued.

This has been a brief review of some of the immediate and long-run
consequences of environmental problems in Thailand. They culminate to

affect human health and reproduction. Environmental hazards threaten life and can cause birth defects. Environmental health hazards are a subject about which we still know relatively little. We have identified only a small proportion of the estimated five million chemicals found in the environment as causing human disease or genetic damage. Some of these, like PCBs, arsenic, DDT, and cadmium are known to be toxic when found in human tissue.[19] What we do not always know is what concentrations of these substances can be "safely" allowed to accumulate. Moreover, we do not know how to eliminate many of these substances completely and we often discover the effects of chemicals only after a severe health problem has been identified for workers in a particular industry or for residents of a specific community.

Causes of Environmental Problems

Modern societies differ from earlier ones in that, through industrialization, they have the means to inflict environmental damage more rapidly and more widely than ever before. High-information and high-energy modern technology has the potential of causing disastrous environmental consequences. In fact, many environmental problems have arisen as unanticipated side effects of technological activities originally intended to improve human life, like the role of the automobile in air pollution or the hazardous effects of radiation to human health. Unfortunately, modern science at present is least developed in its ability to predict consequences and best developed in its ability to create them. Today, members of various advanced industrial societies are well aware and concerned about this irreversible stage of development.

Besides technology, what are the causes that contribute to the environmental problems in Thailand? I have collected data on environmental perception from a representative sample of government officials, industrialists, and the educated public to investigate the social side of environmental problems. I will analyze the data from the cultural, institutional, and individual levels. At all three levels, basically it is individuals who make environmental related decisions—whether as members of a cultural group, which is reflected by their cultural world view; as representatives of institutional systems as government policy makers, politicians, industrialists, or businessmen whose influence make a difference in the society's development trend; or as individual human beings who, although limited by culture and institutions, can become more aware of the environmental consequences of their lifestyle and can make choices, especially political ones, to help save their society from further environmental deterioration. Hence, it logically follows that the environmental perception and behavior of these people will reflect the state of the environmental problems and the potential for improvement of the environmental problems of their society.

Research has shown that there are enormous variations in people's perceptions and evaluations of their environments,[20] particularly in the area of hazard perception,[21] which explains to quite an extent the effectiveness or failure of environmental improvement. For example, Sewell has shown that engineers tend to perceive water pollution as an economic problem, whereas public health officials perceived it as a health problem.[22] In this regard, the way one views the environment is a function of what one does in it. So, unless shared environmental perception and shared concerted effort to fight environmental problems exist, there is little hope for remedying the environmental problems.

As a pilot project to investigate the environmental cognition of the educated Thai in both business and government circles, a checklist of some daily direct and immediate environmental impact was developed and administered to a limited sample from the two groups. To ensure sincere responses, the checklist also consists of "fill-in" items devised to disguise the true purpose of evaluating the respondents' environmental behavior. Actual habitual practices of various environmental related behavior were assessed as well as perception of the environmental relevance of this behavior. The results relating to environmental cognition and behavior of this pilot study are presented in Table 13.5. The analysis below is based on these results and other related studies about the Thai people.

The Cultural Level

At the cultural level, environmental problems are related to our ideas about environmental relationships (our world view, religious beliefs, and scientific thought about the environment) and our technologies (the tools and techniques by which we modify the physical world for our own purposes). In the case of world views, cultures in which humans are believed to be part of the earth rather than its masters, have a tendency towards ecological conservatism. Such cultures, in which earth and nature are seen as parents or sacred forces to be respected and not destroyed more than is necessary for subsistence, promote human responsibility for the physical world. Western industrial societies have been dominated by a world view of humans mastering nature. The productive and reproductive successes of these societies has in turn resulted in the serious environmental consequences with which we are confronted.

TABLE 13.5 Environmental Cognition and Behavior of Some Thai Government Officials and Businessmen

	Areas of Environmental Impact (%)	Habitual Practices Not Known	Perceived Environmental Relevance (%)
Atmosphere			
Smoking	64.6	...	9.2
Burn leaves/grass clipping at home	18.5	...	3.1
Tend to use high powered cars	93.8	...	3.1
Household trash eventually incinerated	7.7	92.3	23.1
Bug sprays, hair sprays, etc.	100.0	...	3.1
Water			
Leave water running when brush teeth and wash dishes	66.2	...	0.0
Use fertilizers, insecticides, drain cleaners	100.0	...	6.2
Use laundry detergents (vs. biodegradable soaps)	100.0	...	0.0
Solid Waste Disposal			
Use plastic bags, styrofoam boxes (preferable as clean and neat)	100.0	...	0.0
Use plastic utensils and cups, paper plates (preferable as clean and disposable)	100.00	...	0.0
Throw away used aluminum can, glass and plastic bottles, plastic cartons, etc.	89.2	...	0.0
Energy			
Amount of gasoline used a year	...	100.0	3.1
Air-conditioned rooms	100.0	...	0.0
Air-conditioner and refrigerator as a percentage of electric bill	...	100.0	0.0

(N=65)

The religious world view of Buddhism teaches living in harmonious alignment with nature and against killing animals and any living thing. In addition, many Thai believe in spirits associated with trees and land—demonstrated by the colorful cloths tied around certain trees and the spirit houses in most house compounds. However, such beliefs and behavior have not stopped Thais (farmers and logging businessmen alike) from destroying seventy percent of their forest in the last twenty-five years. While poor farmers might clear the forest to provide land for housing and their crops, unscrupulous businessmen are driven purely by the motive to make profits by all means. This is why illegal logging has been going on for decades, and is getting worse in response to high market prices.

Even after the government issued a decree banning logging in response to the "flooding" tragedy in southern Thailand, deforestation continues. For example, farmers are paid to cut down trees and immediately transform the logs into hurriedly put up houses, which are then torn down and sold as used wood to evade the law. Such houses are known as *"ban daed diew"* (one-sunned-house).

Besides the deforestation problem, the selling of fertile farmlands for resorts and golf courses, the killing and selling of wildlife, as well as the destroying of coral reefs for businesses, are further evidence of the extent to which Thai cognition is influenced by the monetary material world. In fact, research findings concerning Thai value systems and behavioral patterns have shown that the Thai place high value on "form" (over "content") and material possessions.[23] The Thai perceive material possessions as symbols of being "modern" and high status. This materialism and commercialization surfaces in every circle, including the religious, and manifests itself as the "spending beyond one's means" syndrome in the lower socio-economic class. The concept of "development" held by government officials contributes to this—development being equated with roads, electricity, refrigerators, motorcycles, and the like.

The Thai's cognitive perception of their environment seems to be limited to a person's own self and household. The environmental impact of this is clearly illustrated by trash disposal behavior. Piles of trash and all kinds of wastes are commonly seen thrown outside houses or in a nearby open space. Likewise, Thai factory owners cannot see the consequences of dumping their untreated industrial wastes into rivers. What is thrown into the river is "gone" from the perceived environment. This view leads to one of the most serious cultural misconceptions of the environment—that things can be thrown away. The Thai are fully susceptible to the "throw-away" culture of modern industrial society in which all kinds of disposable products are invented, used, and thrown away. This situation is compounded by a lack of frugality. The "throw-away" culture and ostentatious lifestyle of the well-to-do, is perceived as "modern" and, thus, is widely adopted by others. It is not surprising to find the responses of the educated sample in my study revealing preferences for practices and goods promoting convenience, comfort, and higher status, with almost no concern for the environmental consciousness.

The Institutional Level

The main institutional causes of environmental problems are: (1) business corporations pursuing profit and neglecting environmental costs and (2) corrupt or ineffective government and political institutions. First, in our business world, the environmental and social costs of production are seen as external to the calculation of the company's profits or losses. Since companies usually view air and water as free goods, they can hardly see the reason to pay for costs incurred to prevent pollution. Although treatment equipment is required by law, many factories escape from installation, while others fail to use the equipment to cut costs. The industrialists single-mindedly aim to maximize profit for themselves, seeing environmental prevention as an unnecessary cost. One can take any example from among recent environmental disasters—the "flooding tragedy," the explosion of chemical wastes at the Port Authority, or the explosion of a liquefied petroleum gas tank truck that burned to death so many people in Bangkok—the basic reasons have always been illegal operations, substandard safety precautions, and corruption and laxness in law enforcement by government officials.

Given the low level of environmental consciousness of the businessmen and government officials sampled (shown in Table 13.5), and given the basic motive of maximizing profit, the Thai social system in which persons and relations are always more important than the system, law, principles, and ideology,[24] the occurrence of all these tragic events is understandable. Unless serious action is taken by the government, there is a high probability that the pattern will continue. The seriousness of the problem is demonstrated by the low level of environmental consciousness closest to the self and by the superficial perception of environmental problems more broadly as seen by the responses to an open-ended question: "What do you think are the environmental problems in Thailand?" The responses (Table 13.6) show that perceptions of environmental problems are either physically oriented—what physically affects the respondent most, like heavy traffic congestion, dirty streets, and polluted rivers—and/or reflects what they read in newspapers.

TABLE 13.6 Perception of Environmental Problems in Thailand

Type of Problem	Percent of Respondents
Deforestation	93.8
Traffic Congestions	89.2
Dirty Rivers, Canals	64.6
Littering in Bangkok, Pattaya, Phuket	47.7
Air Pollution	10.8

N=65

The types of products being produced and the technologies which produce them are also causes of environmental problems. The rapid expansion of industrial production after World War II has been accompanied by a shift in technology which is far more polluting than the prewar productive technology.[25] Some examples of this are the shift from biodegradable soaps to phosphate polluting laundry detergents, and the growth in the use of packaging materials for commercial products, especially plastics and nonreusable containers. Consumers have shifted from wood, natural fibers (cotton, wool, and linen), and unpackaged groceries to plastics, synthetics, and packaged and processed foodstuffs. Each of these new products requires more energy and generates more hazardous byproducts than the one it replaced.[26] Take plastics for example, dangerous chemicals are used in their production and their disposal (i.e., PVC when burned releases toxic dioxins into the atmosphere). Environmentalists maintain that the harm plastics do outweighs their worth and have been campaigning against their use. Countries like Switzerland, Germany, Denmark, the Netherlands, Austria, and Italy have taken measures against the production and use of plastics ranging from imposing ecological taxes to boycotting and banning certain products. In Thailand, with its open market economy and emphasis on rapid industrialization, the hazards generated by the unlimited and uncontrolled use of chemical products and the uncontrolled management of chemical wastes are ignored or at best played down by the private and public sectors. Moreover, the public is not well informed and even few educated businessmen and government officials are aware of the environmental and health costs of using plastics.

Inadequate provision for waste disposal is also a major cause of environmental problems. Most industrial waste treatment facilities are either not installed or not operated, largely as a result of lax law enforcement, bribery, and the selfishness of businessmen. Politicians and government officials are also to be blamed. It is ridiculous that Bangkok, the capital city with over eight million people and the location of most industries, still does not have even one wastewater-treatment center when the technology has long been available to make even the most polluted water drinkable. One result of this is the rising rate of morbidity of the top three diseases: diarrhea, dysentery, and food poisoning. In addition, solid waste disposal is managed by dumps and landfills that are in the middle of the heavily congested and densely populated community of "Soi Ornnuj" on Sukhumvit 77 road. On these hills of contaminated waste is a major slum where people live and make a living from searching for sellable items through the trash.

This brings us to the last but also most important institutional cause of environmental problems—that is, the government machinery and political institutions. In most industrial societies, an increasing number of laws and regulations have been created to deal with the social and environmental costs of production. For example, Switzerland and Austria ban the use of those disposable goods that have a short useful lifetime, and Italy has imposed an ecological tax on production of plastics and an additional sales tax on every non-biodegradable plastic bag bought. In Thailand, however, this has, by and large, not been the case. The government is dominated by those who determine and change regulations to fit their own vested interests, ignoring calls for curbing corruption and caring little for environmental protection. Politicians who have bought their way into office seek to recoup their losses after the election. Such a political system, with its rampant corruption and profit orientation, has served to promote the worsening of environmental problems in the country by fostering poor management of the environment, inadequate implementation of existing policies, and lax law enforcement.

In terms of policy formulation, environmental issues have, in fact, been incorporated into government planning since the second national development plan (1967-1971), which included a few clauses concerning forestry protection and conservation of natural resources, with a corresponding budget allocation. The government enacted the Environmental Quality Act of 1975 and created the National Environment Board to serve as its planning and coordinating agency. Responsibility for the administration and control of environmental quality rests with many different ministries and agencies. Projection of environmental quality was explicitly singled out as separate category in the fourth NESD Plan (1977-1981). Thus, there has been no shortage of environmental policies.[27]

Policy is one thing, implementation is another. The distance from words to practices in the Thai context has been especially great. Failure in the implementation of environmental policies can be seen as resulting from the insincerity of the country's decision makers, the inefficiency of government agencies, and widespread corruption. While the National Environment Board (NEB) has no direct administrative authority for managing the environment, the agencies with such authority are numerous and their actions are largely uncoordinated, with each perceiving environmental problems as only part of their other wider responsibilities. Moreover, these agencies have no authority to take legal action against violators. For example, there are about twenty government agencies with responsibilities related to problems of water pollution. The common practice used in response to these problems is to set up additional agencies or committees and to draft additional ineffective regulations.

These administrative shortcomings culminate in the most important factor inhibiting serious action on behalf of the environment—law enforcement. The Thai government has been impotent in enforcing anti-pollution laws. In general, law enforcement has been crippled by the social system that values person and connections over principles and law. Law enforcement has often been selectively applied to those who lack connections or do not have the money to erase their wrongdoings.[28] This problem of laxness of law enforcement in solving environmental problems was highlighted at a recent international conference in Bangkok. After the permanent secretary of the Ministry of Industry presented a detailed survey of the environmental problems caused by 600 chemical plants, their output of hazardous waste increasing from 1.1 million tons in 1986 to an estimated 5.9 million tons by the year 2000, and declared that the ministry was trying to solve these problems by increasing the number of regulations for industries to follow, Tanin Kraivixien, former prime minister and present deputy director of the Chulabhorn Research Institute (CRI), noted that "In Thailand, enforcement is a problem."[29] Then, the question is how can it be ensured that regulations will be followed. Strict enforcement of laws needs dedicated and incorruptible officials as well as businessmen who have a moral conscience. From this perspective, the picture appears to be quite pessimistic. In fact, many foreign companies and businessmen see Thailand as a polluter's haven because of its lack of enforcement of anti-pollution regulations. A good illustration is provided by the practice of dumping tons of toxic chemical wastes from foreign countries at the Port Authority area, this went on for years, until the waste exploded accidentally in 1990.

It is this lack of social control over industrial economic activities that represents our most serious environmental problem, for only our political institutions can make the public decisions required in order to solve environmental problems.

Individual Level

Individuals are left with limited choices within their cultural and institutional context. However, we can still contribute in many ways to help solve environmental problems through our daily actions, by becoming more aware of the environmental consequences of our lifestyles and our consumption patterns, and by making choices that are as environmentally sound as possible. We may also make political choices that help to deal with pressing environmental problems. Unfortunately, we are hampered by Thailand's less than mature democratic political institutions and by the lack of environmental awareness of the majority of Thai people. In regard to the second point, concerned scholars and the media can play an important role by seeking to educate the public and putting pressure on the government.

Notes

1. S. Komin, *Social Dimensions of Industrialization in Thailand* (Bangkok: Research Center, NIDA, 1989).
2. *Bangkok Post*, 8 February, 1989.
3. *Bangkok Post*, 21 June, 1989.
4. *Bangkok Post*, 17 November, 1989.
5. *Matichon*, 21 February, 1989.
6. *Bangkok Post*, 5 September, 1988.
7. *Matichon*, 23 March, 1989.
8. See, for example, TDRI, *Thailand Natural Resources Profile* (Bangkok: TDRI, 1988); S. Bhumibhamon, *National State of the Environment Report* (Bangkok: UNEP, 1988).
9. See *Bangkok Post*, 10 July, 1989.
10. TDRI, *Thailand Natural Resources Profile*, 1988.
11. See V. Tosuwanchinda, *A Report on the Occupational Safety and Health Situation in Thailand* (Bangkok: ILO/PIACT, 1985); P. Hasle, *et. al.*, *Survey of Working Conditions and Environment in Small Scale Enterprises in Thailand*. Technical Report 12 (Bangkok: ILO/NICE Project, 1986).
12. *Bangkok Post*, 16 March, 1989.
13. *Bangkok Post*, 22 November, 1988.
14. TDRI, *Thailand Natural Resources Profile*, 1988.
15. TDRI, *Thailand Natural Resources Profile* (Bangkok: TDRI, 1986).
16. *Matichon*, 27 October, 1989.
17. Tosuwanchinda, *A Report on the Occupational Safety and Health Situation in Thailand*.
18. Department of Health (Bangkok, 1983).

19. R. Bourne and J. Levin, *Social Problems: Causes, Consequences and Interventions* (New York: West Publishing Co., 1983).

20. C. Levy-Leboyer, *Psychology and the Environment* (Beverly Hills: Sage, 1982).

21. W.W. Ittelson, *Environment and Cognition* (New York: Seminar Press, 1973).

22. W.R.D. Sewell, "Environmental perceptions and attitudes of engineers and public health officials," in *Environment and Behavior*, Vol. 3, 1971, pp. 23-60.

23. S. Komin, "Culture and work related values in Thai organizations," paper presented at the International Symposium on Social Values and Effective Organization: Indigenous Experiences in Developing and Newly Industrialized Countries, Taipei, 26-30 November, 1988; *The Psychology of the Thai People: Values and Behavioral Patterns* (Bangkok: Research Center, NIDA, 1991).

24. Komin, *The Psychology of the Thai People.*

25. See B. Commoner, *The Closing Circle* (New York: Bantam, 1971).

26. See R. Bourne, and J. Levin, *Social Problems.*

27. See K. Snidvongs and K. Panpiemras, "Environment and development planning in Thailand," in *Environment and Development in Asia and the Pacific: Experiences and Prospects* (Nairobi: UNEP, 1982), pp. 347-360.

28. Komin, *The Psychology of the Thai People.*

29. *Bangkok Post*, 2 September, 1990.

Selected References

Ahmed, Iftikhar. 1988. "Pro-poor potential," in RIS, *Biotechnology Revolution and the Third World: Challenges and Policy Options*. Pp. 134-149. New Delhi: Research and Information System for Non-aligned and Other Developing Countries.

Amriah Buang. 1982. "The irrelevancy of agricultural innovations." Paper presented to UNESCO regional workshop on "Socio-Cultural Change in Communities resulting from Economic Development and Technological Progress." Bangi, Malaysia, 4-6 October.

Anderson, Robert S., and Walter Huber. 1988. *The Hour of the Fox: Tropical Forests, the World Bank, and Indigenous People in Central India*. New Delhi: Vistaar Publications.

Aparicio, Teresa. 1988. "Philippines: Organisation and models of indigenous ethnodevelopment." *IWGIA Newsletter* 55/56: 79-98.

Arbhanhirama, Anat, *et al*. 1988. *Thailand Natural Resources Profile*. Singapore: Oxford University Press.

Barlow, C. 1978. *The Natural Rubber Industry*. Kuala Lumpur: Oxford University Press.

Bauer, P.T. 1946. "The working of rubber regulation." *Economic Journal* (September): 391-413.

Bautista, Germelino M. "The forestry crisis in the Philippines: Nature, causes and issues." Unpublished paper.

Bawagan, Aleli. 1989. "Laguna Lake for whom?" *Philippine Currents*, October.

Belena, Abe P. "The systematic destruction of the fragile cordillera," *Sarilakas Grassroots Development* 4 (1/2).

Bhumibhamon, S. 1988. *National State of the Environment Report: Thailand*. Bangkok: UNEP.

Booth, A. 1986. "Agricultural taxation: A survey of issues," in P. Shome, ed., *Fiscal Issues in Southeast Asia*: 117-140. Singapore: Institute of Southeast Asian Studies.

Braake, Alex L. ter. 1944. *Mining in the Netherlands East Indies*. Bulletin 4. New York: The Netherlands and Netherlands Indies Council of the Institute of Pacific Relations.

Briones, Leonor M. 1990. "The Philippine external debt: International cooperation toward a people-centered debt strategy." Paper presented to the US-Asia Institute, Philippine International Convention Center, 14 May.

_____. "The continuing debt crisis and the destruction of the environment: Two aspects of the same coin." Unpublished paper.

Brower, Barbara. 1991. *Sherpas of Khumbu: People, Livestock and Landscape*. New Delhi: Oxford University Press.

Budiarjo, Carmel. 1984. *West Papua: The Obliteration of a People*. London: TAPOL.

Burgess, Michael. 1992. "Dangers of environmental extremism: Analysis of debate over India's social forestry programme." *Economic and Political Weekly*. 3 October: 2196-2199.

Buttel, Frederick H., and Martin Kenney. 1988. "Prospects and strategies for overcoming dependence," in RIS, *Biotechnology Revolution and the Third World: Challenges and Policy Options*. Pp. 315-348. New Delhi: Research and Information System for Non-aligned and Other Developing Countries.

Cain, Stephen R., and W.A. Kerr. 1987. "China's changing development strategy: The case of rural electrification." *Canadian Journal of Development Studies* 8 (1): 81-96.

Carino, Joanna K. 1986. "Philippines: National minorities and development." *IWGIA Newsletter* 45: 195-219.

Chaturvari, Mahesh. 1979. "'Second best' technology as first choice." *Ambio* 8: 71-81.

Colchester, M. 1986. "Unity and diversity: Indonesian policy towards tribal peoples." *The Ecologist* 16(2/3).

David, Karina Constantino. 1989. "Access of the poor to basic social services," in *Poverty and Growth in the Philippines*. Manila: Friedrich Ebert Stiftung.

Dembo, David, and Ward Morehouse. 1987. *Trends in Biotechnology Development and Transfer*. IPCT. 32, Technology Trends Series No. 6. Vienna: UNIDO.

_____, Clarence Dias, and Ward Morehouse. 1989. "The vital nexus in biotechnology: The relationship between research and production and its implications for Latin America." Caracas, Venezuela. (mimeo).

Devalle, Susana B.C. 1977. *La Palabra de la Tierra: Protesta Campesina en India, Siglo XIX*. Mexico: El Colegio de Mexico.

_____. 1992. *The Dialectics of Cultural Struggle: Discourses of Ethnicity. Jharkhand*. Delhi: Sage.

Dhawan, B.D., (ed.). 1991. *The Big Dams: Claims and Counter Claims*. New Delhi: Commonwealth Publishers.

Dittmer, Lowell. 1990. "China in 1989: The crisis of incomplete reform." *Asian Survey* 10 (1): 1-25.

Donner, Wolf. 1982. *The Five Faces of Thailand: An Economic Geography*. St.Lucia: University of Queensland Press.

_____. 1987. *Land Use and Environment in Indonesia*. Honolulu: University of Hawaii Press.

Dore, Ronald. 1986. *Flexible Rigidities*. Stanford, CA: Stanford University Press.

Evans, Martin C., *et al.* 1979. *Sector Paper on Agriculture and Rural Development*. Manila: Asian Development Bank.

Expert Task Force. 1983. *The Malaysian Natural Rubber Industry: Report of the Task Force of Experts*. Kuala Lumpur: Malaysian Rubber Research and Development Board.

Fearnside, Phillip M. 1988. "China's Three Gorges Dam: Fatal project or step towards modernization." *World Development* 16 (5): 615-630.

Fernandes, Walter, Geeta Menon, and Philip Viegas. 1988. *Forests, Environment and Tribal Economy: Deforestation, Impoverishment and Marginalisation in Orissa.* New Delhi: Indian Social Institute.

Foljanty-Jost, Gesine. 1988. *Kommunale Umweltpolitik in Japan: Alternativen zur Rechtformlichen Steuerung.* Hamburg: Institut für Asienkunde.

Fujioka, Masao. 1988. *Report and Recommendations of the President to the Board of Directors on the Proposed Loans and Technical Assistance to the Republic of the Philippines for a Forestry Sector Program.* Manila: Asian Development Bank.

Galtung, J. 1989. *Peace and Development in the Pacific Hemisphere.* Honolulu: University of Hawaii Institute for Peace.

Ganjanaphan, Anan. "Community forestry management in northern Thailand," in Amara Pongsapich, Michael C. Howard, and Jacques Amyot, eds., *Regional Development and Change in Southeast Asia in the 1990s.* Pp. 75-84. Bangkok: Chulalongkorn University Social Research Institute, 1992.

Gault-Williams, Malcolm. 1990. "Strangers in their own land." *Cultural Survival Quarterly* 14 (4): 43-48.

Glaeser, Bernhard (ed.). 1983. *Umweltpolitik in China.* Bochum: Brockmeyer.

Goodland, Robert. 1988. "Tribal peoples and economic development: The human ecological dimension," in John Bodley, ed., *Tribal Peoples and Development Issues: A Global Overview.* Pp. 391-405. Mountain View, CA: Mayfield.

Gordon, Alec. 1988. "How many rubber smallholders?." *The Planter* (Kuala Lumpur) 64: 69-75.

_____, and Napat Sirisambhand. 1987. *The Situation of Women Rubber Smallholders in Southeast Asia.* Bangkok: Chulalongkorn University Social Research Institute.

Guerrero, Milagros C. 1967. *A Survey of Japanese Trade and Investment in the Philippines, 1900-1941.* Quezon City: University of the Philippines.

Hardjono, J.M. 1977. *Transmigration in Indonesia.* Kuala Lumpur: Oxford University Press.

Hasle, P., et. al. 1986. *Survey of Working Conditions and Environment in Small Scale Enterprises in Thailand.* Technical Report 12. Bangkok: ILO/NICE Project.

Hicks, George L., and Geoffrey McNicoll. 1971. *Trade and Growth in the Philippines: An Open Dual Economy.* New York: Cornell University Press.

Homjun, Chakrit, et al. 1989. *Impact of Eucalyptus Plantation on Soil Properties and Subsequent Cropping om Northeast Thailand.* Report prepared with the support of USAID and RDI of Khon Kaen University.

Hong, Yip Yat. 1969. *The Development of the Tin Mining Industry of Malaya.* Kuala Lumpur: Oxford University Press.

Howard, Michael C. 1988. *The Impact of the International Mining Industry.* Sydney: University of Sydney, Transnational Corporations Research Project.

_____. 1991. *Mining, Politics, and Development in the South Pacific.* Boulder: Westview Press.

_____. 1992. "Ethnicity, development, and the state in Southeast Asia and the Pacific," in A. Pongsapich, M.C. Howard, and J. Amyot, eds., *Regional Development and Change in Southeast Asia in the 1990s.* Pp. 70-84. Bangkok: Chulalongkorn University Social Research Institute.

Huang Yuanjun, and Zhao Zhongxing. 1987. "Environmental pollution and control measures in China." *Ambio* 16: 473-479.

Hugo, Graeme J.,Terence H. Hull, Valerie J. Hull, and Gavin W. Jones. 1987. *The Demographic Dimension in Indonesian Development*. Singapore: Oxford University Press.

Hunter, A. 1968. "Minerals in Indonesia." *Bulletin of Indonesian Economic Studies* 11: 73-89.

Jacob, J. 1984. "Development of uneconomic tea and rubber lands," in W. Gooneratne and D. Wesumperuma, eds., *Plantation Agriculture in Sri Lanka*. Pp. 183-205. Bangkok: ILO/ARTEP.

Junne, Gerd. 1987. "Bottlenecks in the diffusion of biotechnology from the research system into developing countries' agriculture." Paper presented at the Fourth European Congress on Biotechnology, Amsterdam, June.

_____. 1988. "Incidence of biotechnology advances on developing countries," in RIS, *Biotechnology Revolution and the Third World: Challenges and Policy Options*. Pp. 193-206. New Delhi: Research and Information System for Non-aligned and Other Developing Countries.

Kambara, Tatsu. 1984. "China's energy development during the readjustment and prospects for the future." *China Quarterly* 100: 762-782.

Kannan, K.P. 1982. "Forestry: Forests for industry's profits." *Economic and Political Weekly* 17 (23): 936-937.

Kasryno, F., N. Pribadi, A. Suryana, and J. Musanif. 1991. "Environmental management in Indonesian agricultural development," in Denizhan Erocal, ed., *Environmental Management in Developing Countries*. Pp. 157-173. Paris: Organization for Economic Cooperation and Development.

Kawata, Masaharu. 1990. "Environmental pollution in Leyte." Unpublished paper.

Kazuhiro, Ueta. "Dilemmas in pollution control policy in contemporary China." *Kyoto University Economic Review* 58 (2): 51-69.

Khoo, Khay-Jin. 1978. "The marketing of smallholder rubber," in Kamal Saleh, ed., *Rural Urban Transformation in Malaysia*. Tokyo: United Nations University.

Kinzelbach, Wolfgang. 1983. "China's energy and the environment." *Environmental Management* 7: 310-319.

Kloppenburg, Jr., Jack, and Daniel Lee Kleinman. 1988. "The genetic resources controversy," in RIS, *Biotechnology Revolution and the Third World: Challenges and Policy Options*. Pp. 279-314. New Delhi: Research and Information System for Non-aligned and Other Developing Countries.

Komin, Suntaree. 1989. *Social Dimensions of Industrialization in Thailand*. Bangkok: National Institute of Development Administration, Research Center.

_____. 1991. *Psychology of the Thai People: Values and Behavioral Patterns*. Bangkok: National Institute of Development Administration, Research Center.

Kulkarni, S. 1983. "Towards a social forest policy." *Economic and Political Weekly* 18 (5 February): 191-196.

Kumar, Nagesh. 1987. "Biotechnology in India." *Development* 4: 51-56.

_____. 1988. "Biotechnology revolution and Third World: An overview," in RIS, *Biotechnology Revolution and the Third World: Challenges and Policy Options*. Pp.

1-30. New Delhi: Research and Information System for Non-aligned and Other Developing Countries.

Li Jinchang, Kong Fangwen, He Naihui, and Lester Ross. 1988. "Price and policy: The keys to revamping China's forestry industry," in Robert Repetto and Malcolm Gillis, eds., *Public Policies and the Misuse of Forest Resources*. New York: Cambridge University Press.

Lichauco, Alejandro. 1982. "The international economic order and the Philippines," in Vivencio R. Jose, ed., *Mortgaging the Future*. Pp. 38-47. Quezon City: Foundation for Nationalist Studies.

Lieberthan, Kenneth, and Michel Oksenberg. 1988. *Policy Making in China: Leaders, Structures and Processes*. Princeton, NJ: Princeton University Press.

Lim Teck-Ghee. 1977. *Peasants and Their Agricultural Economy in Colonial Malaya, 1874-1941*. Kuala Lumpur: Oxford University Press.

Luk Shiu Hung, and Joseph Whitney. 1988. "Editor's introduction." *Chinese Geography and Environment* 1 (4): 3-25.

Ma Hong. 1986. "Strive to improve our country's environmental protection work." *Chinese Law and Government* 19 (1): 12-29.

Magallona, Edwin. 1984. "Pesticide use in the banana and sugar industries." *Philippine Labor Review* 8 (1): 14-21.

Manarungsan, Sompop. 1989. *Economic Development of Thailand, 1850-1950: Response to the Challenge of the World Economy*. Bangkok: Chulalongkorn University, Institute of Asian Studies.

Marschall, Wolfgang. 1968. "Metelurgie und Frühe Besiedlungsgeschichte Indonesiens" *Ethnologia* (Köln), 4 (n.s.): 29-263.

McAndrew, John P. 1983. *The Impact of Corporate Mining on Local Philippine Communities*. Manila: ARC Publications.

McDowell, Mark A. 1989. "Development and the environment in ASEAN." *Pacific Affairs* 62 (3): 307-329.

Mey, Wolfgang (ed.). 1984. *Genocide in the Chittagong Hill Tracts, Bangladesh*. Copenhagen: International Work Group for Indigenous Affairs.

Mongkolsmai, Dow, Wachareeya Tosanguan, and Chayan Tantiwasdakarn. *Pulp and Paper Industries in Thailand*. Report prepared by Pacmar, Inc., in association with A&R Consultants Co., Ltd., 1987.

Mooney, Pat Roy. 1988. "Biotechnology and the North-South conflict," in RIS, *Biotechnology Revolution and the Third World: Challenges and Policy Options*. Pp. 243-278. New Delhi: Research and Information System for Non-aligned and Other Developing Countries.

Myint, Hla. 1971. *Southeast Asia's Economy: Development Policies for the 1970s*. Middlesex, England: Penguin Books.

Nadkarni, M.V., K.N. Ninan, and Syed Ajmal Pasha. 1992. "Social forestry projects in Karnataka: Economic and financial viability." *Economic and Political Weekly*, 27 June: pp. A65-A74.

Nartsupha, Chatthip, and Suthy Prasartset (eds.). 1978. *The Political Economy of Siam 1851-1910*. Pp. 205-227. Bangkok: The Social Science Association of Thailand.

Nectoux, Francois, and Yoichi Kuroda. 1989. *Timber from the South Seas.* World-Wide Fund for Nature International Publication.

Nietschmann, Bernard. 1986. "Economic development by invasion of indigenous nations: Cases from Indonesia and Bangladesh." *Cultural Survival Quarterly* 10 (2): 2-12.

_____, and Thomas J. Eley. 1987. "Indonesia's silent genocide against Papuan independence." *Cultural Survival Quarterly* 11 (1): 75-78.

Ofreneo, Rene E. 1980. *Capitalism in Philippine Agriculture.* Quezon City: Foundation for Nationalist Studies.

_____. 1982. "Modernizing the agrarian sector," in Vivencio R. Jose, ed., *Mortgaging the Future.* Pp. 98-127. Quezon City: Foundation for Nationalist Studies.

Omoto, Shimpei, 1981. "Copper smelting and refining." Paper presented at the eighth Joint Conference of the Japan-Philippine Economic Cooperation Committee, March, Fukuoka.

Otten, Mariel. 1986. *Transmigrasi: Myths and Realities: Indonesian Resettlement Policy, 1965-1985.* Copenhagen: International Work Group for Indigenous Affairs.

Palmer, Ingrid. 1978. *The Indonesian Economy Since 1965: A Case Study of Political Economy.* London: Frank Cass.

Panayotoy, Theodore. 1983. "Natural resource management for economic development: Lessons from Southeast Asian experience with special reference to Thailand," in M. bin Yusof and I.H. Omar, eds., *Natural Resource Management in Developing Countries.* Pp. 19-46. Serdang, Selangor: University of Agriculture Malaysia, Faculty of Resource Economics and Agribusiness.

Panchamukhi, V.R., and Nagesh Kumar. 1988. "Impact on commodity exports," in RIS, *Biotechnology Revolution and the Third World: Challenges and Policy Options.* Pp. 207-224. New Delhi: Research and Information System for Non-aligned and Other Developing Countries.

Pannel, Clifton W., and Joseph Ma. 1983. *China: The Geography of Development and Modernization.* London: Edward Arnold.

Pardesi, G. 1980. "Gua incident: Operation annihilation?" *Mainstream* 19 (11): 6-7.

Pascua, Alfredo R. "A hard look on our ills and tribulations." *Sarilakas Grassroots Development* 4 (1/2).

Penz, G. Peter. "Development refugees and distributive justice: Indigenous peoples, land, and the developmentalist state" *Public Affairs Quarterly,* forthcoming.

Poole, Peter J. 1988. "China threatened by Japan's old pollution strategies." *Far Eastern Economic Review,* 23 June: 78-79.

Prasartset, Suthy. 1976. "The tin industry as a non-peasant export production of Thailand," in Carl A. Trocki, ed., *The Emerging Modern States: Thailand and Japan.* Pp. 117-130. Bangkok: Chulalongkorn University, Institute of Asian Studies.

Rawski, Thomas. 1980. *China's Transition to Industrialism.* Ann Arbor, MI: University of Michigan Press.

RIS. 1988. *Biotechnology Revolution and the Third World: An Annotated Bibliography.* New Delhi: Research and Information System for Non-aligned and Other Developing Countries.

Roff, Sue. 1986. Multilateral development assistance destabilizes West Irian."
Cultural Survival Quarterly 10 (1): 38-41.

Ross, Lester. 1984. "The implication of environmental policy in China."
Administration and Society 15 (4): 489-516.

_____. 1988. *Environmental Policy in China*. Bloomington, IN: Indiana University
Press.

Roy, S.B. 1992. "Forest protection committees in West Bengal," *Economic and
Political Weekly*, 18 July: 1528-1530.

Sachs, Ignacy. 1989. *Resources, Employment and Development Financing: Producing
without Destroying—The Case of Brazil*. RIS Occasional Paper No. 30, New Delhi:
Research and Information System for Non-aligned and Other Developing
Countries.

Sagise, P. 1984. "Plant succession and agroecosystem management," in A.T.
Rambo and P. Sagise, eds., *An Introduction to Human Ecology Research on
Agricultural Systems in Southeast Asia*. Los Baños: University of the Philippines
Press, Los Baños.

Sasson, Albert. 1988. "Promise in agriculture: Food and energy," in RIS,
Biotechnology Revolution and the Third World: Challenges and Policy Options. Pp.
55-88. New Delhi: Research and Information System for Non-aligned and
Other Developing Countries.

Scott, Margaret. 1988. "Activists who fight clean." *Far Eastern Economic Review*, 25
February: 44-45.

Sengupta, N. (ed.). 1982. *Fourth World Dynamics: Jharkhand*. Delhi: Authors Guild.

Shamsul, Bahrin, and Husin Ali. 1985. "Challenges facing the smallholder." *Ilmu
Masyrakat* 8: 76-82.

Shira, Vandana. *Ecology and the Politics of Survival: Conflicts Over Natural Resources
in India*. New Delhi: Sage and Tokyo: United Nations University Press, 1991.

Shultz, Theodore W., Kazushi Ohkawa, *et al*. 1968. *Asian Agricultural Survey*.
Tokyo: University of Tokyo Press.

Siddiq, Toufiq A., and Zheng Ching Xian. 1984. "Ambient air quality standards in
China." *Environmental Management* 8 (6): 473-479.

Sirisambhand, Napat, and Alec Gordon. 1992. *Gender Roles and the Situation of
Rubber Smallholders in Thailand*. Bangkok: Chulalongkorn University Social
Research Institute.

Smil, Vaclav. 1979. "A technological future for China." *Ambio* 8: 94-101.

_____, and William Knowland. 1980. *Energy in the Developing World*. Oxford:
Oxford University Press.

_____. 1984. *The Bad Earth* . Armonk, NY: M.E.Sharp.

_____. 1987. "China and Japan in the new energy era," in Peter Nemetz, ed., *The
Pacific Rim: Investment, Development and Trade*. Vancouver: University of British
Columbia Press.

_____. 1988. *Energy in China's Modernization: Advances and Limitations*. Armonk,
NY: M.E. Sharp.

Snidvongs, K., and K. Panpiemras. 1982. Environment and development planning in Thailand, in *Environment and Development in Asia and the Pacific: Experiences and Prospects*. Pp. 347-360. Nairobi: UNEP.

Soesastro, Hadi, and Budi Sudarsono. 1989. "Mineral and energy development in Indonesia," in B. McKern and P. Koomsup, eds., *The Minerals Industries of ASEAN and Australia: Problems and Prospects*. Pp. 198-199. Sydney: Allen & Unwin.

Songanok, Rachanee. 1990. "Commercial eucalyptus plantation in the province of Cha-Choengsao." *Journal of Agricultural Economic Research* (Office of Agricultural Economics, Ministry of Agriculture and Co-operatives, Bangkok). 12 (36).

Sun Jingzhi, *et al*. 1981. *Economic Geography of China*. Hong Kong: Oxford University Press.

Swaminathan, M.S. 1982. "Biotechnology research and Third World agriculture." *Science* 218 (December) 967-972.

_____. 1988. "Biotechnology and sustainable agriculture," in RIS, *Biotechnology Revolution and the Third World: Challenges and Policy Options*. Pp. 33-34. New Delhi: Research and Information System for Non-aligned and Other Developing Countries.

Taylor, Robert. 1981. *Rural Energy Development in China*. Baltimore: Johns Hopkins University Press.

Thailand Development Research Institution. 1989. "Area for fast-growing tree: Whether it is available—an example of using geographic information system to locate the suitable area for fast-growing tree plantation." Paper No. 5 presented to a seminar on "Economic Forest: Reality or Dream," Bangkok, 23 February.

Thukral, Enakshi Ganguly, (ed.). 1992. *Big Dams, Displaced People: Rivers of Sorrow, Rivers of Change*. New Delhi: Sage, 1992.

Tosuwanchinda, V. 1985. *A Report on Occupational Safety and Health Situation in Thailand*. Bangkok: ILO/PIACT.

Tsuda, Mamoru. 1978. *A Preliminary Study of Japanese-Filipino Joint Ventures*. Quezon City: Foundation for Nationalist Studies.

Valentin, Paul. 1987. "The state of the environment in the Cordillera." Unpublished paper.

Vergara, Emma S. 1978. "A second look at the log export ban." *Conservation Circular* (Los Banos: U.P. College of Forestry) 14 (7).

Wihtol, Robert. 1988. *The Asian Development Bank and Rural Developments*. London: Macmillan Press.

Wirkus, Winifred. 1974. History of the Mining Industry in the Philippines: 1898-1941. Ph.D. thesis, Cornell University.

Wirtschafter, Robert M., and Ed Shih. 1990. "Decentralization of China's electricity sector: Is small beautiful?" *World Development* 18 (4): 505-512.

Wong, Lin Ken. 1965. *The Malayan Tin Industry to 1914*. Tucson: University of Arizona Press.

World Bank. 1985. *Report on Indonesia*. Washington, DC: The World Bank.

World Resources Institute. 1989. "The Java garden," in *World Resources 1988-89*: 56-59. New York: Basic Books.

Xin Dingguo. 1988. "The present and long-term energy strategy of China," in James P. Dorian and David G. Findley, eds., *China's Energy and Mineral Industries: Current Perspectives*. Pp. 43-54. Boulder: Westview.

Zamora, P.M. 1977. "An assessment of the environmental effects of mineral extraction and processing." Paper presented at the first National Mines Research Congress, Baguio City, March.

About the Contributors

Susana B.C. Devalle, Professor, Centro de Estudios de Asia y Africa, El Colegio de Mexico, Mexico, DF, Mexico. She is the author of *La Palabra de la Tierra* (1977) and *The Dialectics of Cultural Struggle* (1992).

Alec Gordon, Chulalongkorn University Social Research Institute, Bangkok, Thailand.

Michael C. Howard, Professor, Department of Sociology & Anthropology, Simon Fraser University, Burnaby, B.C., Canada. He is the author of *Political Change in a Mayan Village in Southern Belize* (1977), *Aboriginal Politics in Southwestern Australia* (1981), (co-author) *The Political Economy of the South Pacific* (1983), (co-author) *The Political Economy of the South Pacific to 1945* (1987), *The Impact of the International Mining Industry on Indigenous Peoples* (1987), *Fiji: Race and Politics in an Island State* (1991), *Mining, Politics, and Development in the South Pacific* (1991), (co-author) *Anthropology: Understanding Human Adaptation* (1992), and *Contemporary Cultural Anthropology* (1993), and the editor of several books.

Suntaree Komin, Fellow, Research Center, National Institute of Development Administration, Bangkok, Thailand. She is the author of *Social Dimensions of Industrialization in Thailand* (1989) and *Psychology of the Thai People* (1991).

Nagesh Kumar, Fellow, United Nations University, Institute for New Technologies, Maastricht, Netherlands.

Mark McDowell, Ph.D. candidate, Political Science Department, University of Toronto, Toronto, ON, Canada.

Rene Ofreneo, Associate Professor, School of Labor and Industrial Relations, University of the Philippines, Diliman, Quezon City, Philippines.

G. Peter Penz, Associate Professor, Faculty of Environmental Studies, York University, North York, ON, Canada.

Rosalinda Pineda-Ofreneo, Assistant Professor, Women and Development Program, College of Social Work and Community Development, University of the Philippines, Diliman, Quezon City, Philippines.

Apichai Puntasen, Associate Professor, Faculty of Economics, Thammasat University, Bangkok, Thailand.

Dong-Ho Shin, Ph.D. candidate, Urban Studies Program, University of British Columbia, Vancouver, B.C., Canada.

Donna Winslow, Assistant Professor, Department of Sociology, University of Ottawa, Ottawa, ON, Canada.

Index

Printed and bound by CPI Group (UK) Ltd, Croydon, CR0 4YY

23/10/2024

01778240-0004